DATE DUE

AUG 6 '75 CR		JUN 22 D U	
AUG 4 1975 C R		JUN 23 1981	
JUL 14 1976		JUN 30 1982	
JUL 8 1976			
JUN 29 1977		JUN 30 1982	
JUN 28 1977 B R		NOV 10 1982	
JUN 28 1977		DEC 8 1982	
		JAN 26 1983 RENEWED	
FEB 13 1980		RENEWED	
MAR 24 1980		MAR 2 1983	
MAY 13 1981		RENEWED	
MAY 13 1981		APR 6 1983	
MAY 13 1981		APR 26 1983	
JUN 17 1981		MAY 25 1989	

DEMCO 38-297

ELECTRONIC TROUBLESHOOTING:
A Manual for
Engineers and Technicians

ELECTRONIC
TROUBLESHOOTING:
A Manual for
Engineers and Technicians

Clyde N. Herrick
San Jose City College

RESTON PUBLISHING COMPANY, INC.
Reston, Virginia 22090
A Prentice-Hall Company

Library of Congress Cataloging in Publication Data

Herrick, Clyde N.
    Electronic troubleshooting.

    1. Electronic apparatus and appliances—Maintenance
and repair.   I.   Title.
TK7870.H43   621.381   73–8500
ISBN 0–87909–249–1

© 1974 by
Reston Publishing Company, Inc.
A Prentice-Hall Company
Box 547
Reston, Virginia 22090

10 9 8 7 6 5 4 3 2

Printed in the United States of America.

PREFACE

With the rapid advance of electronic technology, particularly in the solid–state area, the need has become apparent for a generalized trouble-shooting book, both for self instruction and for classroom use. Theory has been held to a minimum in the text, and the troubleshooting aspects of electronic circuitry have been emphasized. A broad spectrum of topics has been chosen for instruction, including AM and FM broadcast radio receivers, hi–fi stereo units, tape recorders, black–and–white television, television cameras, color television, amateur, CB, and mobile radio transmitters and receivers, electronic organs, digital computers, electronic instruments, and MATV/CATV systems. In addition, the first chapter provides a general survey of troubleshooting procedures, and the second chapter describes necessary test equipment, tools, and auxiliary items.

Solid–state circuit analysis and testing is stressed throughout, in accordance with the prevailing state of the art. Topical treatment is instrument oriented, inasmuch as efficient troubleshooting of present–day circuitry demands the use of both basic and comparatively sophisticated test instruments. Accordingly, the oscilloscope and vectorscope are given as much attention as the voltmeter and ohmmeter. Harmonic–distortion meters and intermodulation analyzers are accorded equal treatment with audio oscillators and square–wave generators. Digital logic probes are explained in as much detail as triggered–sweep oscilloscopes. The emphasis throughout the text is directed to instrument applications and interpretation of test results.

It is assumed that the student has completed courses in basic electricity, electronics, and radio theory. It is desirable that courses in black–and–white and color–TV theory either have been completed, or are being taken concurrently with study of this text. It will also be helpful if the student has taken a previous course in electrical and electronic instruments and measurements. However, an alert student can assimilate this text satisfactorily with the superficial knowledge of instruments gained from previous courses in physics and basic electricity and electronics. Mathematics has been held to a minimum, and graphical treatments of quantitative considerations have been utilized in this book. Nevertheless, the student should have completed courses in arithmetic, algebra, geometry, and elementary trigonometry. The author is indebted to numerous manufacturers, as credited throughout the text, for illustrative and technical data. In particular, Howard W. Sams & Co., leading publisher of electronic service data, has been most cooperative in granting permission to reproduce various circuit diagrams.

etry, and elementary trigonometry.

This book is the outcome of many years of teaching experience, both on the part of the author and of his associates at San Jose City College. In a significant sense, this text represents a team effort, and the author gratefully acknowledges the constructive criticisms and contributions of the City College teaching staff. It is appropriate that this work be dedicated to the instructors and students of our technical schools and junior colleges.

CLYDE N. HERRICK

CONTENTS

1

SURVEY OF TROUBLESHOOTING PROCEDURES

1–1 BASIC APPROACH

Certain ordering concepts apply to the basic troubleshooting approach for any unit of electronic equipment. However, there is no single procedure to be followed in troubleshooting the broad spectrum of equipment with which the modern technician must contend. For example, a preliminary evaluation of trouble symptoms is required when starting to troubleshoot a stereo hi–fi receiving system or a digital computer. On the other hand, the testing procedures and the analytical methods that are employed differ considerably in these examples. There are general procedures that apply to individual classifications of electronic equipment. For example, Fig. 1–1 shows the general procedure to be followed in troubleshooting a black-and-white or color–TV receiver. However, procedural details necessarily vary; thus, tests of chroma–circuit action require the use of color signal generators which are not employed in tests of black-and-white circuit action.

1

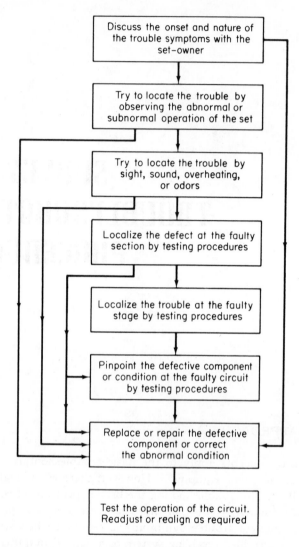

Figure 1–1 General procedure for troubleshooting a black/white or color–TV receiver.

1–2 FUNCTIONAL ORGANIZATION

Trouble symptoms result from equipment malfunction. Accordingly, it is essential for the troubleshooter to have a good understanding of functional organization and circuit action. Functional organization is usually

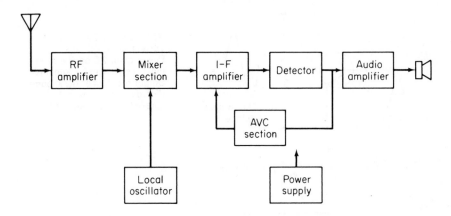

Figure 1–2 A block diagram for a superheterodyne radio receiver.

·shown by block diagrams, as exemplified in **Fig. 1–2**. Note that a functional block, such as the **IF** amplifier section, may contain more than one stage. A block diagram is also a basic signal flow chart; thus, the arrows in **Fig. 1–2** indicate the progress of the incoming signal through the various receiver sections.

Some of the blocks depicted in **Fig. 1–2** provide an amplifying function. Other blocks provide oscillator, demodulator, rectifier, and filter functions. Accordingly, it is necessary for the troubleshooter to understand numerous types of circuit action. Figure 1–3 summarizes the basic transistor circuit actions that are utilized in amplifier circuitry. Figure 1–4 exemplifies the principle of mixer or heterodyne circuit action. It is evident that unless the troubleshooter has a thorough grasp of circuit action, he will be unable to reason from effect to cause, and he will be unable to ascertain which circuit malfunction is causing the observed trouble symptom.

1–3 LOGICAL ANALYSIS

It follows that logical thinking is essential to all troubleshooting procedures. For example, with reference to **Fig. 1–5**, consider a trouble symptom that entails progressive weakening and drop–out of stations as the receiver is tuned from the low–frequency end of the band to the high–frequency end. We would conclude first that the oscillator function in the converter stage is marginal. In turn, there are several possible component defects. The logic employed in this situation is based upon either

3

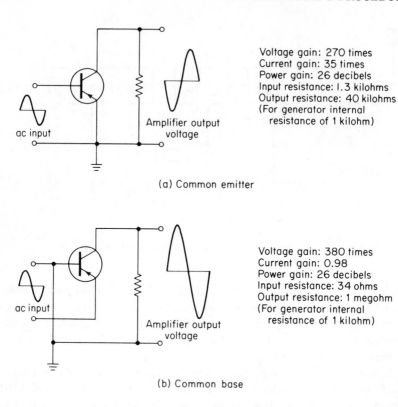

Voltage gain: 270 times
Current gain: 35 times
Power gain: 26 decibels
Input resistance: 1.3 kilohms
Output resistance: 40 kilohms
(For generator internal
   resistance of 1 kilohm)

(a) Common emitter

Voltage gain: 380 times
Current gain: 0.98
Power gain: 26 decibels
Input resistance: 34 ohms
Output resistance: 1 megohm
(For generator internal
   resistance of 1 kilohm)

(b) Common base

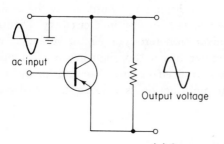

Voltage gain: 1
Current gain: 36 times
Power gain: 15 decibels
Input resistance: 350 kilohms
Output resistance: 500 ohms
(For generator internal
   resistance of 1 kilohm)

(c) Common collector

Figure 1–3 Basic transistor circuit actions.

experience or statistical data. Thus, if Q2 is an economy–type plastic–encapsulated transistor, it would be reasonable to suspect that the transistor is defective. On the other hand, if Q2 is a high–grade transistor, it would be logical to suspect that one of the fixed capacitors in the oscillator circuit has become leaky. Note that although an off–value emitter

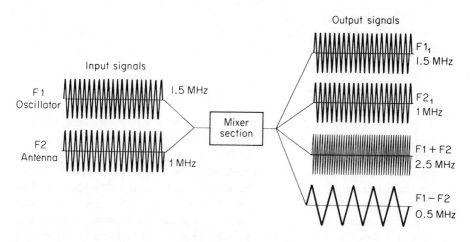

Figure 1–4 Basic mixer or heterodyne circuit action.

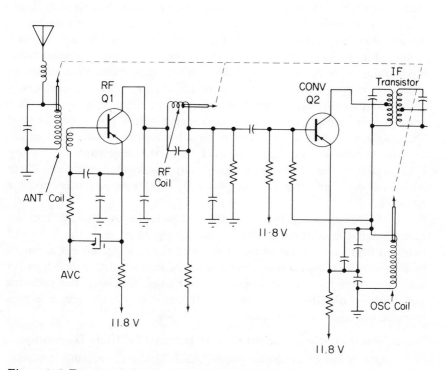

Figure 1–5 Front–end circuitry for an automobile radio.

or base resistor might be impairing oscillator circuit action, this is a less probable cause of the trouble symptom. Other possibilities, such as coil defects or poor connections, are even less probable from a statistical viewpoint.

1–4 NECESSITY FOR SCHEMATIC DIAGRAMS

After a trouble symptom has been localized in a particular receiver section, the block diagram must be supplemented by a schematic diagram. Basic schematic diagrams such as the one depicted in Fig. 1–5 show the components and circuit connections. In some cases, these data are adequate. For example, if the troubleshooter concludes that the most probable cause of the given trouble symptom is a defective capacitor in the oscillator section, he can use a basic schematic diagram to identify the possible suspects. A basic schematic diagram shows how many capacitors are involved, and how each capacitor is connected to key terminals, such as transistor and coil terminals.

On the other hand, a basic schematic diagram is insufficient in many trouble situations. When the cause of a trouble symptom is not clear-cut, and has various possibilities, all of which have approximately the same probability, detailed electrical measurements and tests are required to pinpoint the defective component or condition. These detailed troubleshooting procedures are dependent on the availability of a complete schematic diagram that specifies normal operating voltages, component values, and signal waveforms. For example, the detailed schematic diagram shown in Fig. 1–6 indicates component values and types, normal DC operating voltages, and normal signal waveforms with their peak-to-peak voltage values. Some detailed schematic diagrams also specify DC voltage values under both signal and no–signal conditions. These additional data are often very helpful when analyzing an elusive trouble situation.

In the example of Fig. 1–6, the DC voltages are specified under normal signal conditions. If there were no signal present, the base and emitter of Q15 would rest at zero volts, and the collector potential would be 9.5 volts. It is instructive to note why the emitter of Q15 develops 0.5 volt bias, why the base develops 0.2 volt bias, and why the collector potential is 7.6 volts under signal conditions. The circuit action in this stage is summarized as follows:

1. Under no–signal conditions, Q15 is biased for Class B operation.
2. Q15 is cut off under no–signal conditions, and the collector voltage will accordingly equal the supply voltage (9.5 v).

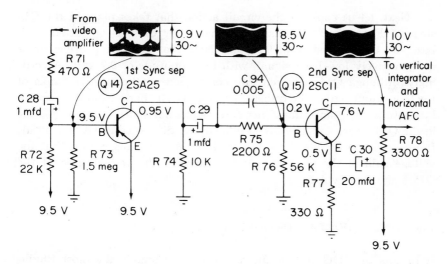

Figure 1–6 A detailed schematic design (*Courtesy of* Howard W. Sams & Co., Inc.).

3. Under signal conditions, the base of Q15 is driven by an AC waveform, and the transistor conducts on the positive excursion of the driving waveform.

4. During the time that Q15 conducts, an IR drop occurs across R78. In turn, this voltage drop subtracts from the supply voltage, and the average potential of the collector is less than the supply voltage value.

5. When Q15 conducts, electrons flow into the emitter terminals of Q15 from C30. Accordingly, there is an average deficiency of electrons on the left–hand electrode of C30, and the emitter of Q15 is positively biased at 0.5 volt.

6. Since Q15 is driven into saturation by the base input signal, some of the supply voltage bleeds into the base circuit on positive peaks of the drive signal. This process leaves an average positive bias of 0.2 volt on the base of Q15.

7. Because of the signal–developed base and emitter bias potentials, Q15 is biased beyond cutoff on the average, and operates in Class C under signal conditions.

8. Although the supply voltage is 9.5 volts, the peak–to–peak voltage of the collector waveform for Q15 is 10 volts. This additional 0.5 volt is provided by the base driving signal during the interval that Q15 goes into saturation.

Next, let us consider the circuit action that takes place in the Q14 stage of Fig. 1–6. The transistor is biased for Class B operation, with base and emitter resting at 9.5 volts. Note that the base is driven by a comparatively low–level signal, with an amplitude of 0.9 volt. Accordingly, although Q14 is driven into conduction on the positive peaks of the base–input signal, the transistor is not driven "hard," and does not go into saturation at any time. Therefore, signal–developed bias is negligible in the Q14 stage, and there is practically no difference in DC potentials from no–signal to signal conditions. It is also for this reason that the voltage gain of the Q14 stage is 9.4 although the voltage gain of the Q15 stage is only 1.2.

1–5 FUNDAMENTAL PROCEDURES

It follows that DC voltage measurements are among the most fundamental troubleshooting procedures. In various situations, the circuit actions that result in particular DC voltage distributions are not entirely simple, and a competent knowledge of circuit action is demanded of the electronic troubleshooter. Note that the supply voltage is generally measured first, because if the supply voltage is incorrect, the DC–voltage distribution in a normally operating circuit will also be incorrect. Most electronics service benches are provided with an adjustable power supply, so that equipment under test can be operated at rated supply voltage. Experienced technicians know that nominal DC voltage values (design–center potentials) are specified in receiver service data, and that certain tolerances are implied. Resistors usually have a tolerance of $\pm10\%$, and may have a tolerance of $\pm20\%$. Transistors are also subject to manufacturing tolerances. Therefore, unless special considerations are involved, voltage values within $\pm20\%$ of nominal specification are regarded as normal. Greater departures from specified values are considered to be probable trouble indications.

Resistance measurements are also among the most fundamental troubleshooting procedures. Out–of–circuit measurement of component resistances is straightforward, and presents few difficulties. On the other hand, in–circuit resistance measurement is generally impractical unless a specialized ohmmeter is employed. As explained in greater detail in Chapter 2, a Hi–Lo FETVM can be used to make accurate in–circuit resistance measurements of solid–state circuitry. When the ohmmeter is operated on its low test–voltage function, semiconductor junctions are not "turned on." In effect, transistors and diodes "look like" open circuits in this situation. For example, if the low test–voltage function is utilized to measure the resistance from the collector of Q14 to ground in Fig. 1–6, a reading of 10k (within a reasonable tolerance) will be obtained unless there is a defective component present.

1–6 SEMICONDUCTOR DEVICE TESTING

Semiconductor devices are tested in two general ways: the device may be disconnected from its circuit, or it may be connected into its circuit. In turn, out–of–circuit test procedures are distinguished from in–circuit test procedures. It is often desirable to make in–circuit tests of diodes and transistors, because it is common practice to solder their terminal leads into printed–circuit boards. Although useful and informative in–circuit tests can generally be obtained, the results are less complete and less accurate than those from out–of–circuit tests. Note that in–circuit tests are usually made to best advantage with conventional instruments, such as VOMs and TVMs. On the other hand, out–of–circuit tests are made to best advantage with specialized semiconductor testers, as explained in greater detail in the next chapter.

Basic semiconductor tests entail measurements of junction resistance, as exemplified in Fig. 1–7. The values of forward resistance and reverse resistance that are measured will depend to some extent upon the test voltage that is applied by the ohmmeter, because junction resistance

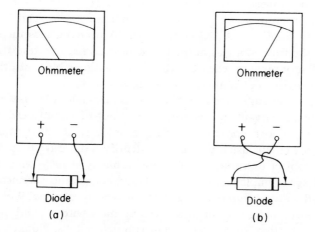

Figure 1–7 Testing a semiconductor diode for front–to–back ratio. (a) Test voltage applied in reference polarity; (b) Test voltage applied in reverse polarity.

is nonlinear. If the forward resistance of a junction is being measured, for example, we might read 30 ohms on the Rx1 range, 80 ohms on the Rx10 range, 300 ohms on the Rx100 range, and 1200 ohms on the Rx1000 range. The reverse resistance of a junction is usually very high, and may not be measurable on the Rx1–meg range of an ohmmeter. If a high forward resistance and/or a low reverse resistance is measured, the front-to-back ratio of the junction is said to be poor, and the semiconductor device is rejected.

The foregoing out–of–circuit test can sometimes be made in–circuit. However, a meaningful in–circuit front-to-back ratio test can be accomplished only when the junction is shunted by a very high value of circuit resistance. If the junction is shunted by a moderate or low value of circuit resistance, one end of the device must be disconnected for the test. Note, however, that other types of in–circuit tests are available as explained in following chapters.

At this point, observe that if a transistor junction is shunted by a moderate or low value of circuit resistance, it is often possible to make "turn–on" and "turn–off" tests, as exemplified in Fig. 1–8. Such tests indicate whether a transistor is workable, or seriously defective. The basis of "turn–on" and "turn–off" tests is as follows: With reference to Fig. 1–8, a DC voltmeter is connected to measure the collector–emitter voltage in the circuit. In (a) and (b), a "turn–off" test is made by short-circuiting the base and emitter terminals of the transistor. If the transistor has normal control action in this test, the voltmeter reading jumps up to the collector supply voltage value when the short–circuit is applied. Otherwise the transistor is defective. Next, in (c), a "turn–on" test is made by connecting a 10k resistor between the collector and base terminals of the transistor. A DC voltmeter is connected across the emitter resistor. If the transistor has normal control action in this test, the voltmeter reading jumps up when the collector voltage is bled into the base circuit. Otherwise the transistor is defective.

Integrated circuits, such as the one depicted in Fig. 1–9, can be tested to better advantage in–circuit than out–of–circuit. DC voltage measurements under normal operating conditions at the terminals of the integrated circuit usually suffice to identify a defective unit. Voltage values are subject to a tolerance of approximately ±20%, when normal supply voltage is employed. In case of doubt, current measurements may also be made. For example, in Fig. 1–9, the lead to terminal 9 of the IC may be opened, and a current measurement made with a VOM or TVM. In normal operation, the current drain is 6 mA ±20%. Note in passing that if C71 or C72 happens to be leaky, the voltage value measured at terminal 9 will be less than 8.4 volts. In such situations, it is essential to

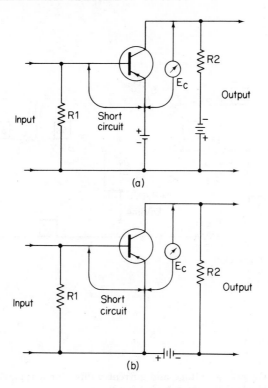

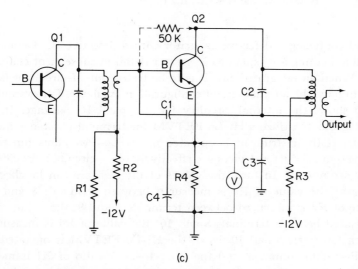

Figure 1–8 In–circuit turn–on and turn–off transistor tests. (a) Turn–off test in
a two–battery circuit; (b) Turn–off test in a one–battery circuit; (c)
Turn–on test in a one–battery circuit.

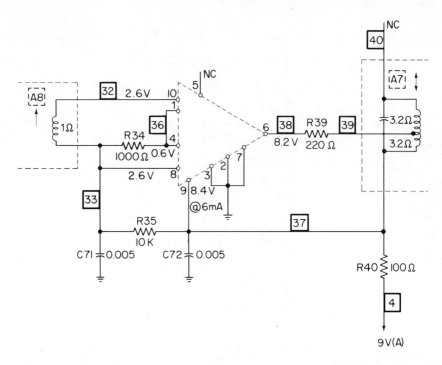

Figure 1–9 Operating voltage and current values for a typical IC (*Courtesy of Howard W. Sams & Co., Inc.*).

avoid confusing a defective capacitor with a defective IC. Details of IC troubleshooting are explained in greater detail in subsequent chapters.

When an integrated circuit is to be tested out–of–circuit, it is essential to have the internal circuit configuration available. For example, Fig. 1–10 shows the internal circuitry for a typical IC package. It is advantageous to utilize a Hi–Lo FETVM for terminal resistance measurements. If the instrument is operated on its Lo–Pwr ohms function, all of the semiconductor devices are effectively open–circuited, provided that they are normal. In turn, the values of the resistors can be checked as follows: the value of R1 is measured between terminals 8 and 9; the value of R2 is measured between terminals 4 and 8; the value of R3 is measured between terminals 8 and 10; the value of R4 is measured between terminals 5 and 10. Next, the Hi–Lo FETVM is operated on its Hi–Pwr ohms function, and the front–to–back ratio of X1 is measured between terminals 3 and 4; the emitter–base junction front–to–back ratio of Q1 is measured between terminals 1 and 2; and the collector–base junction front–to–back ratio of Q3 is measured between terminals

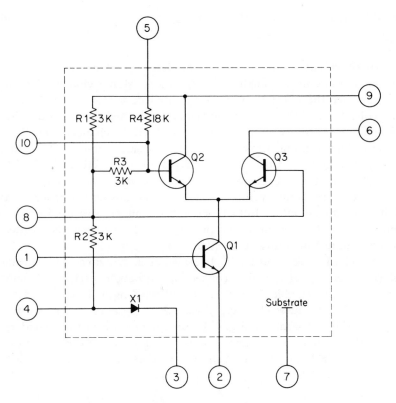

Figure 1–10 A Motorola integrated amplifier package.

6 and 8. The remaining junctions can be checked by means of a comparison test with a similar IC that is known to be good.

1–7 COMPONENT TESTING

Most troubleshooting procedures involve various in–circuit tests of components. That is, a component is never disconnected for test, unless absolutely necessary. As noted previously, when a trouble symptom is localized in a particular stage or circuit, DC voltage measurements are ordinarily made at various points in the circuitry. Out–of–tolerance values indicate that a fault is present, and in most cases a defective component is at fault. For example, if a trouble symptom has been localized in the second sync separator stage in **Fig.** 1–6, and the collector voltage is subnormal, we would suspect that C29 is leaky. Although Q15

could be defective, this possibility is less likely. Therefore, we would disconnect one lead of C29 for further test with an out–of–circuit capacitor
checker.

In other situations, a component defect produces no DC voltage
change in a circuit, although it results in a resistance change. For example, consider the effect of a short–circuited capacitor, C_p in Fig. 1–11.
The collector voltage of Q2 is evidently unaffected. On the other hand,
the winding resistance of the primary in T2 will be changed from 7 ohms
to zero. This winding resistance can be measured in–circuit (with the
supply voltage turned off). It is preferable to use a Hi–Lo FETVM, so
that Q2 cannot affect the resistance measurement. It is important to
note that the resistance value of a coil winding has no necessary relation
to its inductance value. Inductance values can be measured only with
impedance bridges or equivalent instruments. Few service shops attempt
to make inductance measurements.

A coil winding occasionally becomes open–circuited; this is usually
the result of a poor terminal connection, although mechanical damage
can also be responsible. Sometimes a coil winding will be burned out if
a circuit defect causes the supply voltage to be applied directly to the
winding. With reference to Fig. 1–11, an open circuit in the secondary of
T2 would result in a resistance reading of infinity between the cathode
terminal of CR2 and ground. This test is made to best advantage with a

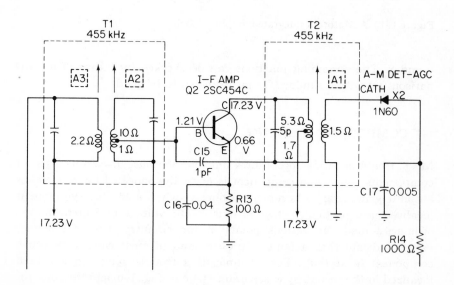

Figure 1–11 Coil winding resistances are usually specified in receiver service data.

Hi–Lo **FETVM**, so that CR2 cannot be "turned on." A less common defect is caused by short–circuited turns in a coil winding. In such a case, the resistance of the winding measures substantially less than the specified value. Leakage between primary and secondary windings of a transformer is another infrequent defect. For example, if leakage occurs between the primary and secondary of **T1** in **Fig. 1–11**, supply voltage will bleed into the base circuit of Q2 and upset the base–emitter bias potential. Since the cause of the incorrect DC voltage distribution is not entirely obvious in such situations, supplementary component tests must be made after the suspected components have been disconnected from their circuit.

1–8 SIGNAL–TRACING PROCEDURES

Signal–tracing procedures are generally employed when there is no signal output from a unit of communications equipment. As an example, consider the superheterodyne radio arrangement depicted in **Fig. 1–12**. In normal operation, voltage waveforms are observable at the input and output of each section in the signal channel, as indicated in the diagram. These waveforms are observed by means of an oscilloscope, such as the one illustrated in **Fig. 1–13**. The test probe of the oscilloscope is applied progressively to the numbered test points in **Fig. 1–12**. Evaluation of screen patterns is as follows:

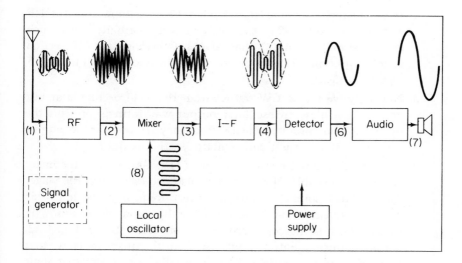

Figure 1–12 Block diagram of a superheterodyne radio receiver.

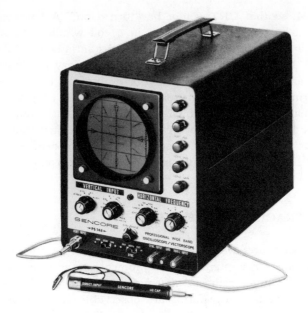

Figure 1–13 A service–type oscilloscope (*Courtesy of* Sencore).

1. If a pattern is observed at test point (1), we know that an input signal is being applied to the receiver from the antenna (or from a generator).

2. If a waveform is found at test point (2), we know that the RF section is workable. Note that the waveform normally has a much greater amplitude at (2) than at (1). Otherwise, the RF section is not amplifying. If there is no output waveform at (1), the RF section is dead.

3. Normally we find a CW output from the local oscillator at test point (3). Unless an oscillator output waveform is present, the mixer will not produce the IF frequency. If there is no output waveform at (3), the local–oscillator section is dead.

4. If we observe a pattern at test point (4), we know that the mixer section is workable. However, it is also necessary that the waveform at test point (4) have a correct frequency (such as 455 kHz). Frequencies can be measured with reasonable accuracy by means of an oscilloscope such as the one illustrated in Fig. 1–13. An off–frequency output waveform from the mixer indicates that a circuit defect is causing the local oscillator to operate at some incorrect frequency.

5. If a waveform is found at test point (5), we know that the **IF** section is workable. Note that the waveform normally has a very much greater amplitude at test point (5) than at test point (4). Otherwise, the **IF** section is not amplifying. If there is no output waveform at test point (5), the **IF** section is dead.

6. Normally we find an audio–frequency waveform at test point (6). This waveform should have approximately the same amplitude as the modulation envelope in the waveform at test point (5). Otherwise, the detector section is defective. If there is no output waveform at test point (6), the detector section is dead.

7. If a pattern is observed at test point (7), we know that the audio section is workable. In normal operation, the waveform amplitude is much greater at test point (7) than at test point (6). If there is no output waveform at test point (7), the audio section is dead.

8. Note that if there is a normal waveform at test point (7), but no sound output from the speaker, the speaker is defective.

1–9 SIGNAL–SUBSTITUTION PROCEDURES

Signal–substitution procedures are preferred to signal–tracing procedures by some technicians. Signal substitution employs the output device of the receiver under test as an indicator. As an example, the speaker of a radio receiver would be utilized as an indicator. With reference to Fig. 1–14, an audio–frequency test voltage would first be injected at test point (1). If there is no audible output from the speaker, we know that the speaker is defective. As noted in the diagram, suitable signals are progressively injected stage by stage, working backward toward the antenna–input terminals. In turn, a seriously defective stage can be localized. Note, however, that weak stages will usually be overlooked in a preliminary test, unless the technician is quite familiar with the receiver performance and with the generator characteristics.

Various types of signal sources can be utilized for signal substitution. The simplest is the pencil–type noise generator, illustrated in Fig. 1–15. It is essentially a solid–state pulse generator. The pulse waveform has a fast rise, and in turn provides strong harmonics. In turn, although the pulse repetition rate occurs at audio frequency, the **IF** and **RF** harmonic content is sufficient to provide signal–substitution tests in the **IF** and **RF** stages of a radio receiver. Note, however, that a simple noise generator is not adequate for a signal–substitution test at (8) in **Fig.**

17

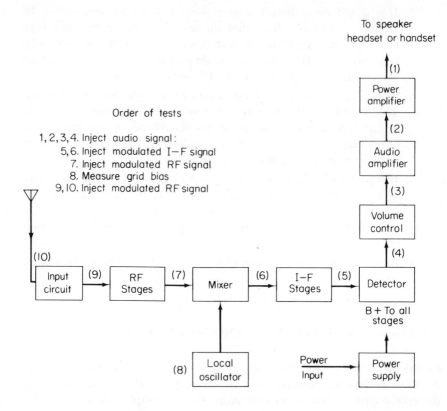

Order of tests

1,2,3,4. Inject audio signal:
    5,6. Inject modulated I—F signal
    7. Inject modulated RF signal
    8. Measure grid bias
   9,10. Inject modulated RF signal

Figure 1–14 General plan of a signal–substitution procedure.

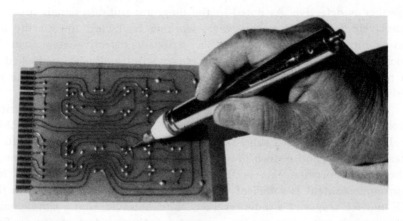

Figure 1–15 A pencil–type noise generator.

1–14. A CW signal of appropriate frequency must be employed to substitute for the local–oscillator output. This requires a service–type RF generator, as described in greater detail in Chapter 2.

1–10 GAIN MEASUREMENTS

Gain is measured in terms of input and output voltage, in terms of dB, or in terms of input voltage and output power. Device gain, stage gain, section gain, and receiver or system gain may be specified and measured in various situations. The gain for a radio receiver, for example, may be specified section by section, as exemplified in Fig. 1–16. Note that the gain of the audio–output stage is specified at 15 times, at a test frequency of 400 Hz. To measure the gain of the audio–output stage, a 400–Hz sine–wave voltage from an audio oscillator would be applied to the base of the transistor. In turn, a TVM would be applied at the collector of the transistor. The output voltage at the collector may be 4 volts rms in a typical situation. Then, if the TVM is connected at the base of the transistor, and an input voltage of 0.26 volt rms is measured, the stage gain would be 15 times.

Comparative voltage, current, and power gain values for basic transistor amplifier configurations are given in Fig. 1–3. Since it is comparatively difficult to measure the current gain of a device or stage, technicians generally measure the voltage gain. Note that the product of the voltage gain and the current gain is equal to the power gain. The power gain is usually expressed in decibels. With reference to Fig. 1–16, the overall gain of the receiver is equal to the product of the individual section gains, or the overall gain is 25,500,000 times. The overall gain might also be expressed, for example, as an output power of 2 watts for an input signal of 500 microvolts. Gain checks will be discussed in greater detail in subsequent chapters.

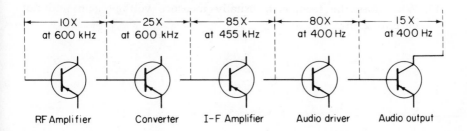

Figure 1–16 Typical voltage gain for each stage of a small a–m receiver.

QUESTIONS

1. Why is it important that a troubleshooter have a good understanding of the functional organization and circuit operation of the equipment under test?

2. What is the purpose of a block diagram?

3. When would a basic schematic diagram of a circuit be insufficient for locating a trouble?

4. What are some of the things that are supplied on a detailed schematic diagram that are not on a basic schematic?

5. What are two of the fundamental measurements used in troubleshooting an electronic equipment?

6. Why is it necessary to use a Hi–Lo FETVM when testing transistor circuits for component resistance?

7. What is the difference between the type of equipment used in testing a transistor in the circuit and out of the circuit?

8. What is the problem in measuring junction resistance with an ohmmeter?

9. How is an ohmmeter used to determine whether a capacitor is leaky?

10. What is the relationship between the inductance of a coil winding and DC resistance of the winding?

11. When would you use signal–tracing procedures on a unit of communication equipment?

12. How is an oscilloscope used to signal–trace a radio receiver?

13. What function does the receiver perform when signal–injection techniques are used for troubleshooting a receiver?

14. In what terms is the gain of a stage or section of a receiver measured?

15. Why does the technician usually measure voltage gain and not current gain?

2

TEST EQUIPMENT, TOOLS, AND AUXILIARY ITEMS

2-1 GENERAL REQUIREMENTS

Most electronic troubleshooting problems require the use of test equipment. Electrical measuring instruments and signal sources are two of the most basic classes of test equipment. A subsidiary class of test equipment includes resistor and capacitor substitution boxes, picture–tube test jigs, substitution speakers and output transformers, variable power supplies, and auxiliary items. Some types of test equipment are qualitative signal sources or indicators, as exemplified by signal tracers and noise generators. That is, a signal tracer indicates the presence or absence of a signal, and provides a rough indication of amplitude. However, a signal tracer is not a true quantitative electrical measuring instrument. Similarly, a noise generator injects an electrical disturbance into a circuit, to determine whether the noise voltage proceeds through the circuit. Although a rough indication of output amplitude is obtained, a noise generator is not a true signal waveform generator. Additional details and considerations of test–equipment characteristics are discussed under subsequent topics.

2-2 VOLT–OHM–MILLIAMMETERS

A volt–ohm–milliammeter (VOM) measures DC voltage, current, and resistance values. AC voltage ranges are also provided by most VOMs. In electronic troubleshooting applications, it is desirable to employ a VOM that has comparatively high sensitivity, in order to minimize indication error due to circuit loading. VOM sensitivity is specified in terms of ohms per volt. For example, the majority of VOMs in current use have a sensitivity rating of 20,000 ohms per volt. Thus, if the VOM is operated on its 50–volt range, its input resistance is 1 megohm. It is also desirable to utilize a VOM that has a comparatively low DC–voltage range, such as 250 millivolts full scale. This feature facilitates the measurement of small bias voltages in solid–state circuitry. Note that the input resistance of a 20,000–ohms–per–volt VOM on its 250–millivolt range is 5000 ohms.

Figure 2-1 illustrates a typical VOM of modern design. One of its practical features is overload protection, which ensures that the meter movement will not be damaged in case excessive input voltage is accidentally applied to the instrument.

Circuit loading is a possible source of measurement error that must be kept in mind when working with voltmeters. For example, in the test

Figure 2-1 A typical VOM suitable for solid–state troubleshooting (*Courtesy of* Simpson Electric Co.).

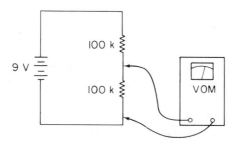

Figure 2–2 This circuit is heavily loaded by a 20,000–ohms–per–volt VOM.

set-up of Fig. 2–2 we would expect to measure 4.5 volts across the lower 100–k resistor. However, if we use a 20,000–ohms–per–volt VOM on its 5–volt range, we will measure 3 volts. This is an error of about 33% caused by circuit loading. That is, the input resistance of the VOM is 100,000 ohms in this example, which reduces the effective value of the 100–k resistor to 50,000 ohms. Thereby, the circuit action is upset, and the anticipated voltage value is not indicated by the meter.

To obtain a more nearly correct measurement in Fig. 2–2, a 100,000–ohms–per–volt VOM may be utilized. In this case, the input resistance of the VOM is 500,000 ohms on its 5–volt range, and the effective value of the lower 100–k resistor becomes 83,333 ohms. In turn, the VOM indicates approximately 4.1 volts instead of 4.5 volts. This is an error of only approximately 9%, due to circuit loading. The advantage of using a VOM with comparatively high input resistance is evident. If a transistor multimeter, covered in Topic 2–3, with an input resistance of 10 megohms on all ranges is utilized, the measurement error in the example of Fig. 2–2 will be negligible.

Another type of circuit loading involves capacitive shunting of the circuit under test by the input capacitance of the voltmeter. This type of loading is particularly severe in high–frequency tuned circuits. For example, if a conventional VOM with open test leads is used to measure the base bias in a VHF oscillator circuit, the oscillating action may stop, or the generated frequency may be far in error. In either case, a serious error is likely to result in the indicated DC voltage value. Therefore, such tests should be made with a voltmeter that has a very small value of input capacitance. Many TVMs have an isolating probe for DC–voltage measurements, consisting of a 1–megohm series resistor in the probe housing. In turn, the effective input capacitance is reduced to about 2 pF, which permits valid DC–voltage measurements in high–frequency oscillator circuits.

2–3 TRANSISTOR MULTIMETERS

Transistor multimeters, also called **TVMs**, field–effect multimeters, or **FETVMs**, have an advantage over VOMs in respect to sensitivity. For example, a typical **FETVM** has an input resistance of 15 megohms on all DC–voltage ranges. In turn its circuit–loading effect is much less than that of a VOM, particularly on low–voltage ranges. Field–effect transistors provide a higher input resistance than the bipolar transistors in a TVM. The instrument illustrated in **Fig. 2–3** features a first DC–voltage range of 0.1 volt, full scale. Its rated accuracy is ±1.5% of full scale value.

Note that instruments in this category measure peak–to–peak AC–voltage values. On the other hand, a VOM can only measure rms AC–voltage values of sine waveforms. The ability to measure peak–to–peak voltages of complex waveforms is often of considerable assistance in troubleshooting procedures.

A unique feature of the instrument illustrated in **Fig. 2–3** is its Hi–Lo ohmmeter function. As noted previously, normal semiconductor junctions do not "turn on," provided the test voltage is sufficiently low, such as 0.08 volt. Therefore, this feature permits measurement of many resistance values in–circuit, and provides more extensive test data than

Figure 2–3 A field-effect transistor multimeter (*Courtesy of* Sencore).

conventional ohmmeters. Note that if a semiconductor junction is short–circuited or leaky, this defect will upset resistance readings even though the applied test voltage is very low. In other words, the defective junction does not "look like" an open circuit, no matter how low the test voltage may be.

2–4 AM AND FM SIGNAL GENERATORS

Amplitude–modulated (AM) signal generators are used in troubleshooting and aligning radio receivers, and for signal–injection tests in television receivers. Numerous types of signal generators are available, from simple RF test oscillators to standard signal generators. A comparatively simple signal generator is illustrated in Fig. 2–4. This type of instrument is widely employed in radio service shops that specialize in AM broadcast receiver work. Although the economy–type signal generator is not highly accurate, and lacks an output–level meter, it serves a useful purpose. This form of signal generator, or RF test oscillator, typically provides a frequency range from 150 kHz to 435 MHz. Frequencies above 30 or 40 MHz are ordinarily not fundamentals, but harmonics. Amplitude modulation at 400 Hz is commonly provided.

Frequency–modulated (FM) signal generators are used in troubleshooting and aligning radio receivers and FM sections of television receivers. FM signal generators for stereo–FM servicing and maintenance are called stereo–FM signal generators, or simulators. A typical stereo–FM signal simulator is illustrated in Fig. 2–5. It provides L and R stereophonic signals at 400, 1000, or 5000 Hz. Monophonic signals are also provided at the same frequencies. Stereo test signals are available

Figure 2–4 A simple signal generator, of the test–oscillator type (*Courtesy of* Heath Co.).

25

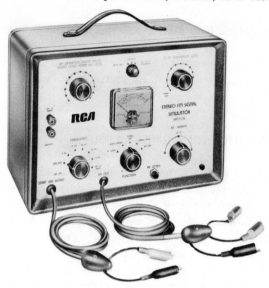

Figure 2–5   Appearance of a stereo–FM signal simulator (*Courtesy of* RCA).

at 19, 38, 67, and 72 kHz. One FM sweep signal at a center frequency of 100 MHz is employed for overall receiver testing. This type of generator is widely used in high–fidelity service shops.

General communications work requires comparatively elaborate and sophisticated signal generators. A typical FM–AM communications generator is illustrated in Fig. 2–6. This instrument has an RF range from 54 to 216 MHz, with an accuracy of ±0.5%. It has a calibrated output system with a range from 0.1μV to 0.2V. All harmonics and spurious outputs are at least 30 dB below the output signal level. RF leakage from the instrument is sufficiently low that measurements can be accurately made at 0.1μV. Amplitude modulation is provided from 0 to 50% at 400 Hz. Frequency modulation is provided with a deviation range from 0 to 250 kHz. Generators in this category are often called standard signal generators.

2–5 SWEEP–SIGNAL GENERATORS

Sweep–signal generators, or sweep–frequency generators, are specialized types of FM generators that are utilized primarily in television alignment procedures. However, sweep generators are also used in troubleshooting procedures, and all–channel sweepers are employed in CATV troubleshooting and maintenance work. The chief distinction between a

Figure 2–6 An FM–AM signal generator for communications service (*Courtesy of* Hewlett–Packard Co.).

television sweep generator and an FM signal generator is in the extent of deviation that is provided. A conventional TV sweep generator has a maximum sweep width of approximately 15 MHz, whereas a typical FM signal generator has a maximum sweep width of about 0.25 MHz. In visual–alignment procedures, technicians often use a TV sweep generator in FM receiver servicing, and operate the generator at reduced sweep width.

A sweep generator must be used with a marker generator in order to identify key frequency points on a frequency–response curve. A marker generator is merely an accurately calibrated signal generator. The majority of present–day sweep generators have built–in marker generators. For example, the sweep–and–marker generator illustrated in Fig. 2–7 provides both FM sweep and CW marker signals. A built–in marker-adder section is also included. This feature maintains markers at uniform amplitude at any point on a frequency–response curve, whether the marker is placed at the top of the curve or down in a sound trap. As explained in greater detail subsequently, a marker adder functions by mixing the marker signal with a sample of the sweep signal, and adding the resulting beat waveform to the swept output waveform from the receiver under test. In turn, the mixed waveform voltages are fed to an oscilloscope. The essential point is that the marker signal is not passed through the receiver.

It is instructive to note the characteristics of the generator shown

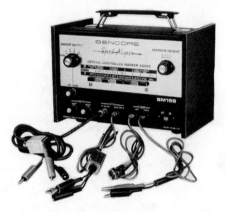

Figure 2–7 A sweep–and–marker generator used in black–and–white and color–TV servicing procedures (*Courtesy of* Sencore).

in Fig. 2–7. Sweep–frequency signals are provided for the **VHF** channels, the **IF** section, and the chroma section. **UHF** channels are covered by means of harmonics from the **VHF** output. An **FM IF** output with a center frequency of 10.7 **MHz** is also available. Crystal–controlled markers at key alignment frequencies are individually or simultaneously displayed, as depicted in **Fig.** 2–8. The output signal level is calibrated over a range from 10μV to 0.1V. A switching arrangement permits the frequency progression to increase from left to right, or from right to left. This feature enables the technician to duplicate the aspect of specified response curves in receiver service data.

One pitfall to be carefully avoided is the premature assumption that a circuit or section needs realignment before adequate troubleshooting

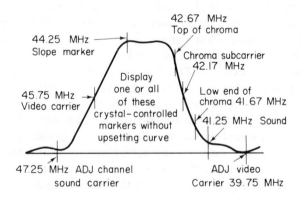

Figure 2–8 Post–injection markers displayed on a frequency-response curve.

tests have been made. Professional electronic technicians make it a rule
to delay all alignment procedures until the last, after any defective components have been replaced. In the great majority of situations, the
apparent need for realignment disappears after the defective component
or components have been found and replaced. Of course, there are exceptions to the general rule, as when it is known that alignment adjustments have been tampered with. After the need for realignment has been
established, it is essential to obtain the applicable alignment procedure
which is included in the receiver service data. If a random "touch–up"
approach is employed, a very poor outcome can be anticipated.

In those cases in which realignment of TV circuits is actually required, a sweep generator is extremely useful. It is instructive to consider a simple tuned circuit, as depicted in Fig. 2–9. We can connect an
RF generator with a constant output to the circuit, a voltmeter with a
detector probe across the coil, and measure the frequencies at which
various voltages occur (the frequency response of the L C circuit). By
starting at a low RF frequency and slowly turning the frequency dial of
the RF generator in small steps, and recording the resulting DC voltage

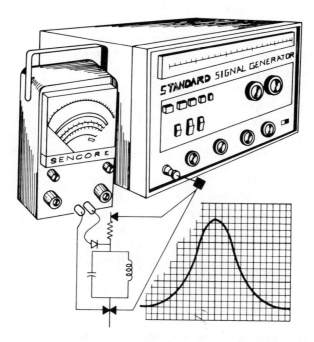

Figure 2–9 Point–by–point method of frequency response–curve plotting.

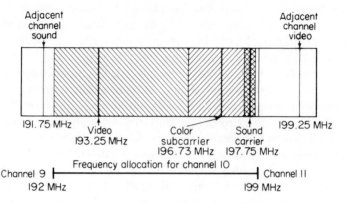

Figure 2–10 Basic spectrum of frequencies transmitted by a TV station.

on a graph, we can plot a response curve of the tuned circuit. In turn, we can determine its resonant frequency and its bandwidth.

However, if each tuned circuit that is adjusted during TV alignment had to be plotted in this manner, a prohibitive amount of time would be wasted. On the other hand, if a sweep generator is employed with an oscilloscope for an indicator, this visual–alignment procedure auto-matically plots the frequency–response curve on the CRT screen, and the result of any alignment adjustment becomes immediately apparent. Of course, this method can be used to view the frequency response of a single LC circuit, of two or more circuits in sequence, of a complete IF section, or of the RF and IF sections in sequence. This is called an overall frequency–response curve. Figure 2–10 shows the basic spectrum of frequencies that are transmitted by a TV station. Figure 2–11 depicts the relation of a typical IF response curve to this frequency spectrum. TV alignment procedures are more extensively detailed in Chapter 6.

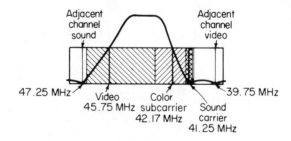

Figure 2–11 Relation of an IF response curve to the spectrum of associated TV frequencies.

2–6 WIDE–BAND SWEEP–FREQUENCY GENERATORS

Various forms of wide–band sweep–frequency generators are utilized in electronics technology. In this topic, a wide–band RF sweep generator is designated as one that sweeps all of the VHF TV channels at the same time. This type of generator is employed principally for checking the frequency–response characteristics of CATV systems, distribution systems, antenna signal boosters, line amplifiers, matching transformers, and other units that operate over the entire VHF spectrum. A typical generator is illustrated in Fig. 2–12. In application, a wide–band sweep–frequency generator is often set at a center frequency of 140 MHz, with a deviation from 55 MHz to 225 MHz. Details of generator application are discussed in a subsequent chapter.

2–7 AUDIO OSCILLATORS

Audio oscillators are utilized in various areas of electronic troubleshooting, and a low–distortion audio oscillator is the basic test instrument in high–fidelity troubleshooting and maintenance. Although no industry standards have been established, it is generally agreed that high–fidelity reproduction entails a frequency response that is uniform within ±1 dB

Figure 2–12 A wide–band sweep–frequency generator that covers the low and high VHF bands (*Courtesy of* Jerrold Electronics Corp.).

31

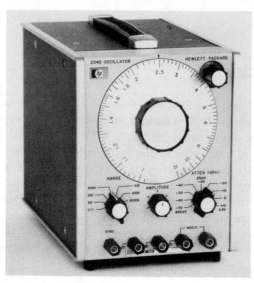

Figure 2–13 A good–quality audio oscillator (*Courtesy of* Hewlett–Packard Co.).

from 20 Hz to 20 kHz, with less than 1% harmonic distortion. In turn, an audio oscillator should have the same frequency range, and less than 1% harmonic distortion. A good–quality audio oscillator is illustrated in Fig. 2–13. This oscillator has a frequency range from 20 Hz to 20 kHz, at a maximum harmonic distortion of 0.5%. It has a maximum output voltage of approximately 40 volts rms and an output impedance of 600 ohms. Dial calibration accuracy is rated at ±1%.

Audio oscillators of this type are designed particularly for amplifier testing, transmission–line measurements, loudspeaker testing, frequency comparison, and audio troubleshooting procedures. Some applications, such as preamplifier testing, require a comparatively low–level signal. The audio oscillator described above has an attenuation range of 40 dB; that is, it has a minimum output level of approximately 0.4 volt. To obtain an output level of 0.04 volt, for example, a 10–to–1 resistive pad would be utilized between the oscillator and the preamp under test. Application methods and requirements are explained in greater detail in following chapters.

2–8 HARMONIC DISTORTION METERS

As noted in the previous topic, high–fidelity reproduction requires a low percentage of harmonic distortion, on the order of less than 1%. Meas-

Figure 2–14 A harmonic distortion meter for high--fidelity service (*Courtesy of* Heath Co.).

urement of harmonic distortion requires the use of a harmonic distortion meter, such as the one illustrated in Fig. 2–14. The instrument operates by trapping the fundamental frequency of the test signal, and indicating the level of any harmonic frequencies that might be present, owing to nonlinearity in the amplifier under test. (See Fig. 2–15.) A range from 20 Hz to 20 kHz is covered by the instrument shown in Fig. 2–14. Percentage of harmonic distortion is indicated by a calibrated meter. In addition, the instrument operates as an audio voltmeter. This function is useful in checking the output level of an amplifier.

2–9 INTERMODULATION DISTORTION METERS

An intermodulation analyzer can be used instead of a harmonic distortion meter to check nonlinear operation in amplifiers or other audio units. However, an IM analyzer is more elaborate than an HD meter, and is less widely utilized in service shops. On the other hand, IM analyzers are employed in all professional and engineering test and design laboratories. An IM analyzer is a two–tone test instrument, whereas an HD meter is a single–tone test instrument. An IM test is made on the basis of beat frequencies, instead of harmonic frequencies. Figure 2–16 illustrates a typical intermodulation analyzer.

Figure 2–17 shows the plan of an IM analyzer. A two–tone signal is utilized, which is applied to the amplifier or unit under test. In turn, the two–tone output signal is processed through a high–pass filter, is rectified, and is then processed through a low–pass filter. Finally, the remain-

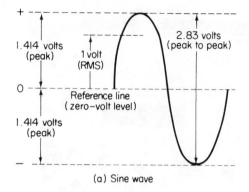

(a) Sine wave

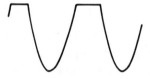

(b) Unsymmetrical clipping generates even harmonics

(c) Symmetrical clipping generates odd harmonics

Figure 2–15 Amplitude distortion is accompanied by generation of harmonic frequencies.

Figure 2–16 A typical intermodulation analyzer (*Courtesy of* Heath Co.).

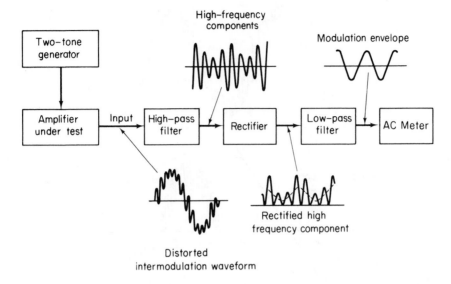

Figure 2–17 Plan of an intermodulation analyzer.

ing AC signal is indicated by a meter. Filter action is employed to separate the IM distortion products (if any) from the incoming signal. That is, if nonlinearity is present in the amplifier under test, a beat signal is generated, which produces an amplitude–modulated waveform. This process is depicted in Fig. 2–18. This AM wave envelope is separated from the complete IM waveform by low–pass/rectifier/high–pass filter action, so that a percentage IM reading is obtained from the calibrated meter.

2–10 SQUARE–WAVE GENERATORS

Square–wave generators are used to determine the transient response of electrical and electronic devices, units, or systems. Figure 2–19 illustrates a high–performance square–wave generator. Transient response is indicated by various distortions in a reproduced square wave which is displayed on the screen of an oscilloscope. Basic forms of square–wave distortion are depicted in Fig. 2–20. This topic is explained in greater detail in subsequent chapters. The generator shown in Fig. 2–19 has a rise time of less than 5 ηs, with a repetition rate from 1 Hz to 10 MHz. A maximum peak–voltage output of 5 volts into 50 ohms is provided, with a minimum peak–voltage output of 0.05 volt. The DC–coupled out-

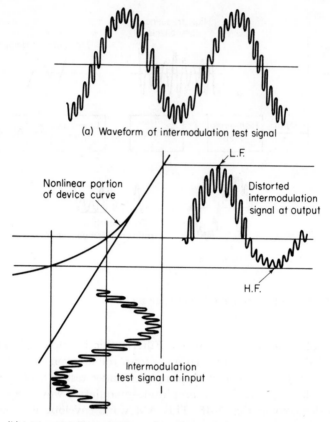

(a) Waveform of intermodulation test signal

(b) Intermodulation waveform distortion resulting from nonlinear amplifier characteristics

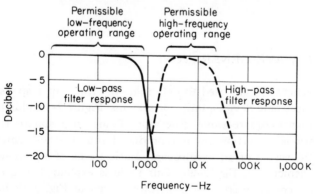

(c) Characteristic of signal filters in a typical intermodulation–distortion analyzer

Figure 2–18 Sequence of IM signal processing.

Figure 2–19 Appearance of a good–quality square–wave generator (*Courtesy of Hewlett–Packard Co.*).

put is a negative–going square–waveform. This feature prevents baseline shift with rep rate changes. A series capacitor can be used to obtain an AC pulse output.

2–11 PULSE GENERATORS

Pulse generators are basically similar to square–wave generators, except that a rectangular waveform is provided. That is, the duty cycle (ratio

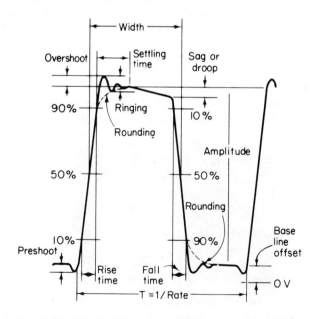

Figure 2–20 Basic forms of square–waveform distortion.

Figure 2–21 A pulse generator suitable for digital–computer circuitry tests (*Courtesy of* Hewlett–Packard Co.).

of "on" and "off" times) of a pulse generator is low, and may be very small. Pulse generators are used in numerous areas of electronic troubleshooting. For example, digital–computer circuitry is analyzed and evaluated to best advantage with a suitable pulse generator and oscilloscope. Some pulse generators are comparatively sophisticated. The generator illustrated in Fig. 2–21 is called a two–channel binary waveform generator, or simply a word generator. It produces variable–length serial digital words at rates up to 10 MHz. Selections of two 16–bit word sequences are available, and a single action combines these two 16–bit words in series to provide a 32–bit word at the output of both channels. A remote programming feature permits rapid conversion of parallel words to serial words. Figure 2–22 depicts typical output pulse waveforms. Miniaturized pulse generators are used to troubleshoot digital equipment, as shown in Fig. 2–23.

2–12 OSCILLOSCOPES

Three general classes of oscilloscopes are utilized in electronic troubleshooting procedures. The most basic characteristic in each class is its

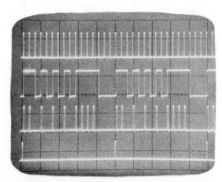

Figure 2–22 Typical pulse waveforms used in digital–computer troubleshooting (*Courtesy of* Hewlett–Packard Co.).

Figure 2–23 Miniaturized pulse generator and signal sensor used in troubleshooting digital equipment. These two logic test devices from
Hewlett–Packard form a stimulus–response test set that is the size
of two pens. One is a mini pulse generator, the other a new test
probe that makes sense out of the signals in logic circuits. Together
or alone they add ease and speed to troubleshooting digital equipment, in design, production or service.

vertical–amplifier frequency response. Thus, the audio–frequency type
of oscilloscope has a vertical–amplifier frequency response from DC (or
a low audio frequency) to at least 20 kHz. Most instruments in this
class have a vertical–amplifier frequency response up to 50 or 100 kHz.
High–performance oscilloscopes feature low–distortion (high–fidelity)
deflection amplifiers. Vectorscope application is usually also provided.

Next, the service–type oscilloscopes designed for black–and–white
television troubleshooting have a vertical–amplifier frequency response
from DC (or a low frequency such as 30 Hz) to at least 1 MHz. Most
oscilloscopes in this class have a vertical–amplifier frequency response
up to 1.5 or 2 MHz. Since they are not intended for high–fidelity tests
and measurements, few of these oscilloscopes employ highly linear deflection amplifiers. On the other hand, lab–type scopes with frequency
response in this range feature low–distortion deflection amplifiers. Some
lab–type scopes also provide identical vertical and horizontal amplifiers,
which permit practical vectorscope application in low–level circuits.

Wide–band oscilloscopes are used in color–TV troubleshooting, and provide vertical–amplifier frequency response to at least 4 MHz. Many scopes in this class have a vertical–amplifier frequency response up to 5 or 10 MHz. The more elaborate instruments feature calibrated and triggered sweeps, to facilitate waveform analysis. Figure 2–24 illustrates an intermediate type of oscilloscope that provides a vertical–amplifier frequency response from DC to 4 MHz, with free–running and several triggered–sweep positions. High–performance triggered–sweep scopes are comparatively costly, and are not widely used in service shops. On the other hand, technicians who specialize in electronic computer troubleshooting always employ high–performance triggered–sweep scopes.

Figure 2–25 illustrates a dual–trace triggered–sweep oscilloscope with provisions for vectorscope application. A dual–trace display of a horizontal–sync pulse with color burst accompanied by an expanded color–burst pattern is shown in Fig. 2–26. A typical vectorgram pattern is illustrated in Fig. 2–27. Vectorscope applications will be explained in some detail in Chapter 8. Note that a vectorscope is always used in conjunction with a color signal generator, described in Topic 2–15. A triggered–sweep oscilloscope has various advantages in television servicing procedures, provided the operator knows how to use the instrument. For example, color–TV stations transmit vertical–interval test signals, as shown in Fig. 2–28. This waveform provides a critical evaluation of the signal–channel response in a television receiver. However, a vertical–interval test signal (VITS) cannot be displayed with a conventional scope—a triggered–sweep scope is required for this type of testing.

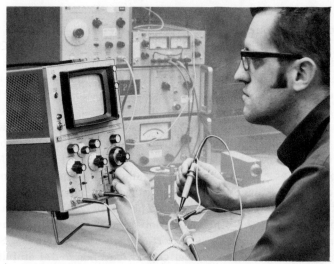

Figure 2–24 An intermediate type of oscilloscope (*Courtesy of* Sencore).

Figure 2–25 A dual–trace triggered–sweep oscilloscope with vectorscope facilities (*Courtesy of* Sencore).

A dual–trace oscilloscope is used to check time relations or time coincidence of various waveforms. This function finds extensive application in troubleshooting high–speed digital circuitry, in which numerous precisely–timed circuit actions take place. A marginal defect in a component can cause a phase shift or time delay which throws associated circuit actions out of time with one another. In this situation, waveforms

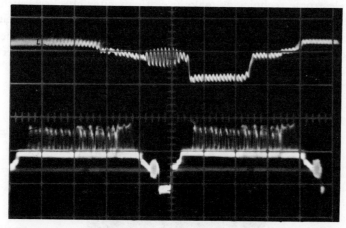

Figure 2–26 A dual–trace display of a horizontal–sync pulse and color burst, accompanied by an expanded color–burst pattern (*Courtesy of* Sencore).

41

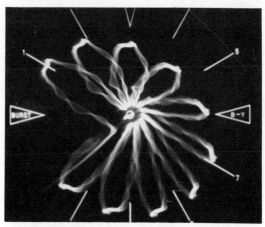

Figure 2–27 Typical vectorgram display (*Courtesy of* Sencore).

can be checked in pairs for proper timing in order to close in on the defective component. This type of testing can also be accomplished with a single–trace oscilloscope, although the procedure is somewhat more involved. That is, external synchronization must be employed, and tests must be made sequentially instead of simultaneously. Of course, triggered sweep is essential in both of the procedures.

2–13 CAPACITOR TESTERS

Most electronic technicians make use of capacitor checkers, such as the one illustrated in Fig. 2–29. This is a bridge–type instrument that measures capacitance values, indicates power–factor values, and checks for leakage. A Wheatstone resistance bridge is included, which serves the

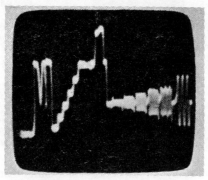

Figure 2–28 Display of a VITS waveform with a triggered–sweep oscilloscope.

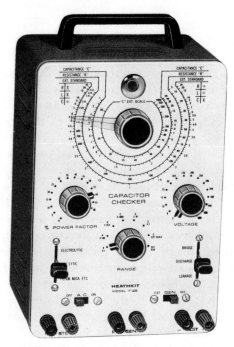

Figure 2–29 Typical service–type capacitor checker (*Courtesy of* Heath Co.).

same basic function as an ohmmeter. Capacitance values are measured out–of–circuit only, with the capacitor connected to the test terminals of the instrument. The range switch is set to a suitable position, and the bridge is balanced by rotating the pointer to the required position on the capacitance scale. Balance is indicated by an eye tube, which opens to maximum at the exact balance setting.

If a capacitor is open–circuited, the eye tube remains closed at all settings of the bridge. If a capacitor is leaky, the eye tube opens to less than maximum, and the capacitance scale indication will be incorrect. This situation is called a "shallow null." Even small amounts of leakage show up clearly on the leakage test. The operator sets the voltage control to the rated working voltage of the capacitor, such as 200 volts, and sets the range switch to its leakage position. If there is marginal leakage in the capacitor, the eye tube will close. An electrolytic capacitor is checked for its power factor, in addition to leakage and capacitance value. That is, the power–factor control must be turned to obtain complete balance of the bridge. In turn, the scale of the power–factor control indicates the power factor of the electrolytic capacitor.

2–14 SEMICONDUCTOR DEVICE TESTERS

Various semiconductor device testers are utilized in electronic trouble-shooting procedures. As noted previously, ohmmeters are employed to check the front–to–back ratios of semiconductor junctions. Voltmeters are used to make in–circuit measurements of electrode potentials and to indicate the result of a "turn on" or "turn off" test. Most transistor testers operate on the basis of gain checks, and are designed for either in–circuit or out-of–circuit application. In addition, junction leakage is measured in out–of–circuit tests. It is impossible to measure leakage in-circuit, but this defect becomes evident as a shift in DC voltage distribution. If leakage is substantial, it also becomes evident as reduced stage gain. Bipolar transistors are checked for current gain (beta value), and field–effect transistors are checked for transconductance (Gm).

A typical in–or–out–of–circuit transistor tester is illustrated in Fig. 2–30. It measures the beta value (or stage gain) of a bipolar transistor by applying an AC current to the base and measuring the resulting AC current from the collector. It measures the transconductance (or stage gain) of a field–effect transistor by applying an AC voltage to the gate, and measuring the resulting AC voltage from the drain. An I_{cbo} leakage

Figure 2–30 An in–or–out–of–circuit bipolar–FET transistor tester (*Courtesy of* Sencore).

test can be made on bipolar transistors out–of–circuit (leakage between collector and base with the emitter open). An I_{gss} leakage test can be made out–of–circuit on field–effect transistors (leakage between gate and source with the source shorted to the drain). Special tests are provided for enhancement–type field–effect transistors, RF transistors, and high–power transistors.

2–15 COLOR SIGNAL GENERATORS

Most color–TV technicians utilize keyed–rainbow color–bar generators. A rainbow signal is a sine–wave voltage with a frequency of 15,750 Hz below (or above) the chroma subcarrier frequency of 3.579545 MHz. Service–type rainbow generators operate at approximately 3.56 MHz. A rainbow signal is also called a sidelock signal, or a linear phase sweep. It is usually keyed by a square–wave modulator into 11 "bursts" followed by a horizontal sync pulse. The first "burst" following the sync pulse operates as a color burst in the receiver color–sync system. Thus, 10 chroma bars are displayed on the screen of a color picture tube. Most service–type generators have RF output only, although a few generators also have video–frequency output for chroma signal–injection procedures. Comparatively elaborate instruments, such as color–TV analyzers, have both RF and video–frequency outputs available.

Figure 2–31 illustrates a typical color signal generator. Like most service–type generators, it also provides white–dot and cross–hatch outputs, which are employed in color picture–tube convergence procedures.

Figure 2–31 A color signal generator with white–dot and cross–hatch outputs (*Courtesy of* Sencore).

The cross–hatch pattern can be switched to display vertical lines only, or horizontal lines only. A single dot and a single cross display are also available. A 4.5 MHz crystal is provided to assist in accurate adjustment of the fine–tuning control in the receiver under test. The RF output frequency range is tunable from channels 2 through 6. This feature ensures that the operator will be able to select an interference–free channel for receiver testing.

As noted previously, color signal generators of the keyed–rainbow type are used with vectorscopes in troubleshooting the chroma–demodulator and matrix sections of color–TV receivers. Test procedures and examples of vectorgram analysis will be discussed in greater detail in Chapter 8.

2–16 HAND TOOLS

Both general and specialized hand tools are required by the electronic technician. Basic necessities include:

1. An assortment of screwdrivers, such as those illustrated in Fig. 2–32.
2. An assortment of Phillips–head screwdrivers, as shown in Fig. 2–33.
3. A set of hex wrenches, as illustrated in Fig. 2–34.
4. An assortment of hex key wrenches and spline key wrenches, as shown in Fig. 2–35.
5. Various types of pliers, including combination, long–nose, standard electrician's pliers, and all–purpose electrician's pliers, as seen in Fig. 2–36.

Figure 2–32 Typical set of screwdrivers used by electronic technicians.

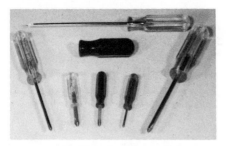

Figure 2–33 An assortment of Phillips–head screwdrivers.

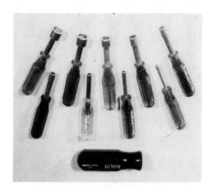

Figure 2–34 A set of hex wrenches, used in electronic servicing procedures.

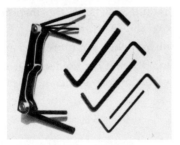

Figure 2–35 Hex and spline key–wrench sets.

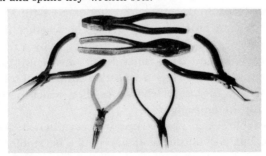

Figure 2–36 Electronic technician's assortment of pliers.

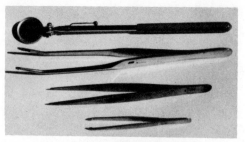

Figure 2–37 Utility tweezers and inspection mirror.

Figure 2–38 A set of box end wrenches.

6. Large and small tweezers are often useful, and a small mirror on an extension handle assists in the inspection of compact electronic units (Fig. 2–37).

7. A set of box end wrenches is required, as exemplified in Fig. 2–38.

8. A soldering gun, pencil–type soldering iron, soldering pick, heat sink, and roll of rosin–core solder are necessary.

2–17 TEST JIGS

Various forms of test jigs are of practical utility in electronic trouble-shooting procedures. For example, a color picture–tube test jig permits the color–TV technician to quickly determine whether improper operation is being caused by a defective picture tube or by a receiver malfunction. Similarly, VHF and UHF tuner repairs are facilitated by the availability of appropriate test jigs. Inspection and repair of various radar chassis is eased by the use of chassis cradles, which permit the chassis to be turned up at any angle.

An adjustable bench power supply is essential; it does not need to be regulated, but the output voltage should be stable and free from drift. A line–isolation transformer is essential for servicing "hot–chassis"

Figure 2–39 Regulated power supply (*Courtesy of* PACO Electronics Co. Inc.).

receivers (**Fig. 2–39**). It "cools" the chassis, so that the technician is not endangered by shock from the 117–volt power line. In addition, a line-isolation transformer minimizes power–supply ground loops, which often cause erroneous instrument readings. For example, it is generally difficult or impossible to make waveform checks in "hot–chassis" circuitry with an oscilloscope, unless a line–isolation transformer is employed.

QUESTIONS

1. What is the purpose of a signal tracer?
2. What is the purpose of a VOM?
3. Why is it desirable to have a high sensitivity factor for a VOM?
4. What is the internal resistance of a 20,000–ohms–per–volt meter on the 250–millivolt range?
5. What is the advantage of the Hi–Lo ohms operation of the TVOM?
6. What are sweep–signal generators?
7. What is the purpose of amplitude–modulated signal generators?
8. What is the chief distinction between a television sweep generator and an FM sweep generator?
9. What are the applications of an audio oscillator?
10. What is the purpose of a harmonic distortion meter?

11. How is percentage of harmonic distortion indicated?
12. What is the basic difference between a harmonic distortion meter and an intermodulation meter?
13. How are square–wave generators used in circuit testing?
14. What is the difference between a square–wave generator and a pulse generator?
15. What are the requirements of an oscilloscope for testing a color television?
16. What measurements can you usually make with a capacitor checker?
17. What type of color signal generator is most often used by TV technicians?
18. Why is the output from some TV–signal generators tunable from channel 2 to channel 6?
19. What is the purpose of the white–dot and cross–hatch outputs from a color signal generator?

3

RADIO RECEIVER
TROUBLESHOOTING

3-1 GENERAL CONSIDERATIONS

Radio receivers may be of the AM, FM, or AM/FM varieties. They may
cover only the broadcast band(s), or they may include one or more
short–wave bands. An occasional broadcast receiver may also provide
a long–wave band. Nevertheless, there are basic troubleshooting principles that apply to any of the above receiver designs. The first step is to
note the trouble symptom(s) that may be involved. Most symptoms can
be classified under "dead" receiver, weak output, distorted sound, incorrect dial calibration, tuning drift, poor selectivity, noisy output, intermittent operation, motorboating, impaired AVC action, or mechanical
defects. A mechanical defect might involve a broken dial cord or damaged speaker cone, for example.

Sometimes preliminary localization of a fault can be made by analyzing the receiver operation. As an illustration, if an AM/FM receiver
is "dead" on its AM function, but operates normally on its FM function,
the technician turns his attention to the AM section of the receiver.
Again, if a multiband receiver operates normally on the broadcast band,
but is "dead" on one or more of the short–wave bands, the technician
turns his attention to the circuitry associated with the "dead" band. Or,

if a receiver operates normally on its battery power source, but does not operate properly on its AC power supply, it is indicated that there is a defect in the power supply.

3-2 ANALYSIS OF WEAK-OUTPUT SYMPTOMS

Weak sound output can be caused by a fault in any receiver section. In the great majority of cases, there is only one fault to be pinpointed. Unless an obvious defect is present, such as a broken antenna lead or a torn speaker cone, sectionalization must be made by means of signal–tracing or signal–injection tests. This procedure will localize the defect to the input circuit, RF stages, local oscillator, mixer, IF section, detector, audio section, or the power supply. An oscilloscope is the most informative signal–tracing instrument. It should have a high–sensitivity vertical amplifier, and a bandwidth of at least 1.5 MHz. To avoid undue circuit loading and detuning of resonant coils, a low–capacitance probe should be utilized.

Signal tracing starts at the antenna–input circuit of the receiver. Figure 3–1 illustrates normal waveforms at the input coupling coil. The peak–to–peak voltage at this point is usually in the order of several thousand microvolts (several millivolts). As depicted in Fig. 3–2, a typical voltage gain of 10 times is normally provided by the RF amplifier stage. That is, the technician expects to see an increase in waveform amplitude of 10 times when he moves the scope test probe from the input to the output of the RF amplifier. (The receiver must be tuned to the applied signal frequency.) If the gain figure is seriously subnormal, he proceeds to look for a circuit defect in this stage. Other-

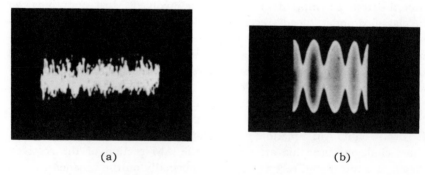

(a)   (b)

Figure 3–1 Normal waveforms across the input coupling coil of a radio receiver. (a) Antenna signal input; (b) AM generator signal input.

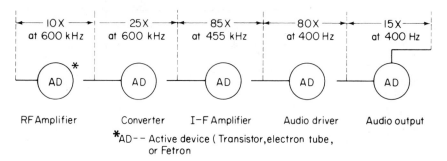

RF Amplifier   Converter   I–F Amplifier   Audio driver   Audio output

*AD-- Active device (Transistor,electron tube,
      or Fetron

Figure 3–2 Typical specified voltage gain values for a small AM radio receiver.

wise, he proceeds to check at the output of the converter (or mixer) stage.

Note that the local–oscillator signal is combined with the **RF**–amplifier signal in the converter stage. Typical frequency relations are depicted in **Fig. 3–3.** In many receivers, the oscillator waveform has a much greater amplitude than that of the antenna signal waveform, unless the latter happens to be unusually large. This means that converter gain must be measured in practice at the input of the first **IF** amplifier stage. Thereby, the tuned coupling circuit between the converter and the IF amplifier serves to filter out the oscillator signal, leaving only the difference frequency between the antenna and oscillator signals. Test points for measurement of converter gain are exemplified in **Fig. 3–4.**

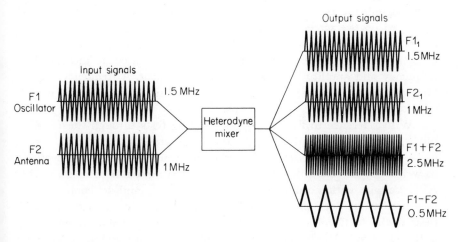

Figure 3–3 Frequency relations in a heterodyne mixer stage.

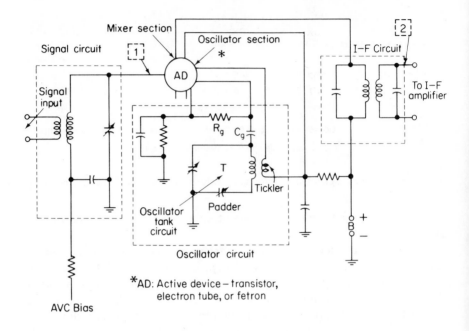

Figure 3–4 Test points for measurement of converter–stage gain.

Notice also in **Fig. 3–3** that the individual waveforms shown in the output circuit cannot be separated by an oscilloscope used with a low-capacitance probe. This limitation applies in similar manner to conventional signal tracers. In a conventional scope test, a complex waveform is displayed on the screen; this complex waveform is the resultant of the component waveforms that are present. On the other hand, a heterodyne frequency meter (wavemeter) can be utilized, if desired, to measure the frequency and amplitude of each signal component. A frequency meter, however, is generally used only in troubleshooting sophisticated types of communication receivers or systems.

With reference to **Fig. 3–2**, signal–tracing tests through the **IF** amplifier, audio driver, and output audio stage are straightforward. Since weak–output symptoms can be caused by trouble in the AVC section, as well as by defects in the signal sections, it is advisable to measure the AVC voltage with normal antenna–signal input, and to compare the measured value with the specified value in the receiver service data. If desired, the AVC voltage can be clamped at a chosen value by means of batteries or an AVC bias box. This clamp (override) voltage is applied between the AVC bus and chassis ground. Since the clamp–voltage source has a low internal resistance, it effectively disables the **AVC** action and establishes the **AVC** voltage at a fixed value.

Analysis of a weak–output symptom can also be made by signal–substitution procedures. As depicted in Fig. 3–5, an **AM** signal generator is used to apply a test signal at the input of each stage or section. A 0.1–μ blocking capacitor is connected in series with the generator out-

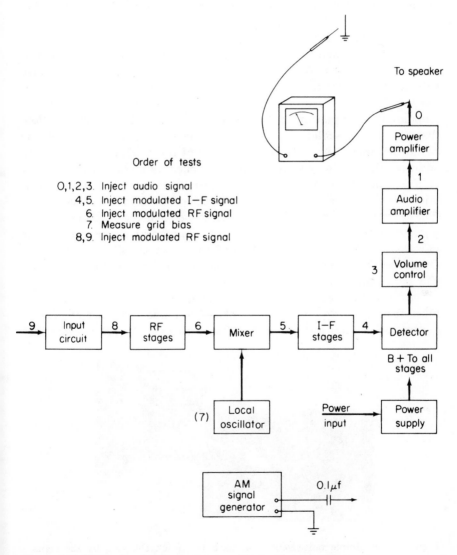

Figure 3–5 Basic signal–substitution troubleshooting procedure.

put lead, to prevent drain–off of bias voltage from the device under test. When the audio–frequency output from the generator is applied at test point 0, for example, a reference reading is noted on the voltmeter, which will be stipulated as V_0. Next, when the same output signal is applied from the generator at test point 1, a higher reading is normally noted on the voltmeter, which will be stipulated as V_1. If the gain figures indicated in Fig. 3–2 were specified for this receiver, the normal ratio V_1/V_0 would be 15. Note that if gain measurements are not of concern, rough signal-injection tests can be made with a simple noise generator, such as the one illustrated in Fig. 3–6.

In this same manner, the gains of the other receiver sections are progressively measured. When the technician encounters a section or stage with subnormal or zero gain, the defect causing the weak–output symptom has been localized. The next step is to close in on the defect and pinpoint the component or condition that is causing the trouble. Note that the local oscillator is usually checked by measurement of the signal–developed bias voltage at the base of the oscillator transistor. In addition, the oscillator section can be checked by signal injection. That is, if a generator signal of correct frequency and suitable amplitude is injected at test point 7 in Fig. 3–5, the receiver will resume normal operation in case of an oscillator defect. Remember that the amplitude of the injected signal from the generator must be progressively reduced as the injection point is moved back toward the input circuit (test point 9). Otherwise the audio section will become overloaded, and meter readings will be meaningless.

After the defective section or stage has been localized, further tests must be made to close in on the fault and to pinpoint the defective component. Sometimes a defect is obvious, such as a charred resistor or a

Figure 3–6 A simple noise generator, used for quick–checking by the signal-injection method.

broken lead. However, it is usually necessary to make systematic voltage and resistance measurements in the defective circuit. As exemplified in Fig. 3–7, normal DC operating voltages are specified in receiver service data. Incorrect DC–voltage readings generally occur in the trouble area. For example, if C14 is short–circuited, the emitter voltage will be zero. Or, if the collector junction of Q1 is leaky, both the base and emitter voltage values will be abnormally high. Note that the emitter voltage of Q1 normally changes to some extent as the receiver tuning control is turned. That is, an oscillator circuit usually has greater output at some frequencies than at other frequencies. If the emitter voltage remains unchanged, the technician suspects an open or leaky transistor, or possibly an open oscillator coil.

When DC–voltage measurements are inconclusive, the technician often proceeds with resistance measurements. Many resistance measurements can be made in–circuit, although component disconnection is sometimes required. As an illustration, the resistance values of R11 and R13 (Fig. 3–7) can be measured in–circuit. It is helpful to use a hi–lo FETVM in this application, because the polarity of the test voltage does not need to be taken into account. By way of comparison, if an ordinary ohmmeter is employed, the test leads must be applied in suitable polarity to reverse–bias the adjacent device junction. Thus, if the value of R11 is to be checked with a conventional ohmmeter, the positive test lead must be applied to the emitter of Q1, inasmuch as this is an NPN type transistor. Note that resistance charts published in service data for solid–state receivers are based on the use of a hi–lo FETVM.

Next, consider a component defect that cannot be pinpointed without disconnections. With reference to Fig. 3–7, a weak–output symptom can be caused by leakage in C2, C17, or C18. Localization is made on the basis of a subnormal resistance reading across any one of the foregoing capacitors. However, since the subnormal reading could be caused by any one of the three capacitors, it becomes necessary to systematically disconnect one lead of each capacitor in turn, to make a conclusive test. On a statistical basis, the technician would suspect C2 first, since this is an electrolytic capacitor. Figure 3–8 shows how capacitors are ordinarily tested in–circuit with a DC voltmeter. With operating voltage applied to one lead of the capacitor, and with the other lead disconnected, even slight leakage in the capacitor will be indicated by a substantial voltage reading on a low–range TVM scale.

Open capacitors can often be pinpointed to best advantage by signal–tracing or signal–injection tests. For example, if C1 is open (Fig. 3–8), an oscilloscope will show the presence of signal at the input end, but no signal at the output end of the capacitor. Again, if C2 is open,

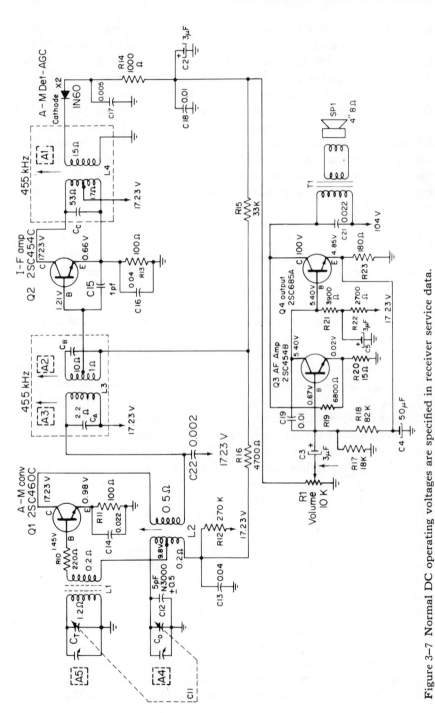

Figure 3-7 Normal DC operating voltages are specified in receiver service data.

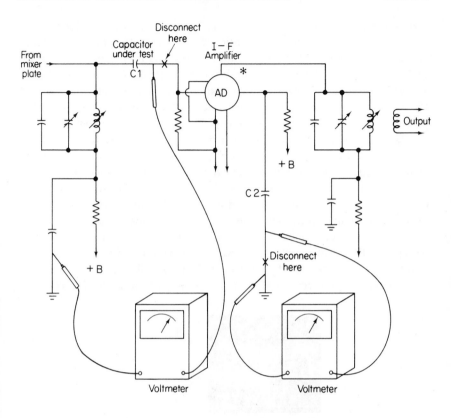

* Active device, such as transistor, Fetron,
  or electron tube.

Figure 3–8 Capacitors are often checked with a voltmeter.

an oscilloscope will show a high signal amplitude at the input end of the
capacitor, whereas a very low signal amplitude is normally found at this
point. A quick test that is often utilized when a capacitor is suspected of
being open is called a "bridging" test; the technician merely connects a
known good capacitor across the suspect, to determine whether normal
circuit operation will resume. Of course, the test capacitor should be of
approximately the same value as the capacitor with the suspected
"open."

3–3 ANALYSIS OF DISTORTED–OUTPUT
SYMPTOMS

Distortion analysis requires the use of a program signal or an AM generator signal. That is, a noise generator is useless in this application, and

even an AM generator has limited usefulness unless an oscilloscope is employed as an indicator. Conventional AM generators provide a single audio tone at 400 Hz (or 1 kHz), and unless this tone signal is substantially distorted, evaluation by ear becomes difficult. On the other hand, an oscilloscope will show 5% distortion clearly, as illustrated in Fig. 3–9. Most technicians "play it by ear" and utilize a program signal when troubleshooting distortion in ordinary AM broadcast receivers.

Overload distortion, exemplified in Fig. 3–9, is most likely to occur in a high–level stage such as Q4 in Fig. 3–7. Again, frequency distortion, in which only part of the audio spectrum is passed, is more likely to occur in the front end or IF sections. For example, an open bypass or decoupling capacitor in the front end or in an IF stage can cause positive feedback (regeneration) with the result that tuning becomes very critical, and the signal sidebands are "cut." In turn, either the low, medium, or high audio frequencies are increased abnormally and the other audio frequencies are attenuated or rejected. This trouble can also be caused by replacing a conventional loop antenna with a high–Q ferrite–rod antenna.

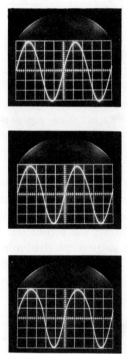

Figure 3–9 Audio sine–wave distortion. (a) 3% harmonic distortion; (b) 5% harmonic distortion; (c) 10% harmonic distortion.

Other possible causes of distorted–output symptoms include: incorrect substitution of a high–beta transistor for a moderate–beta transistor, loose or missing coil shields, leaky coupling capacitors, off–value bias or load resistors, subnormal power–supply voltage, or a replacement speaker with a seriously incorrect impedance rating. When a push–pull audio–output stage is present, weak and distorted output is very likely to be caused by a defective transistor in the push–pull circuit. An oscilloscope will show which half of the stage is defective, even though DC–voltage measurements may be inconclusive. Analysis is more difficult when an output transformer is present, because of interaction through the transformer windings. Accordingly, disconnection tests may be required to pinpoint the defective component.

3–4 ANALYSIS OF NO–OUTPUT SYMPTOMS

A no–output trouble symptom may or may not be accompanied by noise (a rushing sound) from the speaker when the volume control is turned to maximum. If a normal noise level is present, it is logical to conclude that the signal sections are operating at normal gain. That is, most of the noise voltage is contributed by the input section of a receiver, since this energy is amplified by the IF and audio sections. On the other hand, a negligible proportion of the noise voltage is contributed by the output section of a receiver. When a normal noise level is present, the technician knows that the trouble will neither be found in the AVC section nor in the power supply. Instead, suspicion falls on the local oscillator. Initial tests are made to determine whether the oscillator is operating, and whether it is operating on–frequency.

In case the oscillator is operating normally, and the noise level is normal, suspicion next falls on the RF input circuit. As an illustration, the antenna tuning capacitor C_T in Fig. 3–7 could be "open" or "shorted," or the primary of L1 could be "open." Systematic troubleshooting of a no–output symptom involves the same basic procedure as in the case of a weak–output symptom, as explained in the previous topic. That is, a marginal component defect can result in a weak–output symptom, whereas total failure of the same component can result in a no–output symptom. For example, a small amount of leakage in a coupling capacitor can produce a weak–output symptom, but a "short" in the same capacitor can produce a no–output symptom. In the latter case, the associated transistor may also be damaged by excessive current flow, and circuit resistors may be burnt.

3-5 ANALYSIS OF POOR-SELECTIVITY
SYMPTOMS

Poor selectivity denotes failure of a receiver to separate stations that are operating on different frequencies, assuming the stations could be separated if the receiver were operating normally. Good selectivity requires that the tuned circuits operate together as a team, to pass the desired signal frequency and to reject other signal frequencies. (See Fig. 3-10.) As an illustration, with reference to Fig. 3-7, tuning capacitors C_T and C_o must "track" properly in order to provide good selectivity. That is, the difference in resonant frequencies between L1 and L2 must be 455 kHz at any point along the tuning dial. Otherwise, when C_T is set to tune in a certain signal frequency, C_o will be heterodyning a different signal frequency. This trouble can be caused by replacing a loop antenna (L1) or an oscillator coil (L2) with an incorrect type. Improper "tracking" can sometimes be eliminated by realignment of the front end. However, if the inductance of a replacement coil is far off value, poor selectivity cannot be corrected by alignment procedures.

Another common cause of poor selectivity is misalignment of the IF

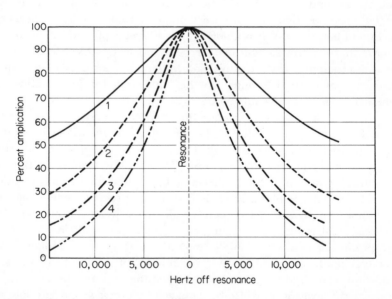

Figure 3-10 Good selectivity results from several tuned circuits operating together as a team.

signal channel. For example, tuned circuits A1, A2, and A3 in Fig. 3–7 must resonate at 455 kHz. Otherwise, poor selectivity and interference will result. Impaired sensitivity also occurs, because most of the gain is contributed by the IF amplifier. This misalignment is often the result of someone's having tampered with receiver adjustments. However, various component defects can cause the same difficulty. As an illustration, if capacitor C_A, C_B, or C_C should "open up," the associated tuned circuit will be thrown off resonance. Occasionally, the technician finds shorted turns in an inductor—a solder splash, for example, can cause this defect. Misaligned stages can be pinpointed by a systematic alignment procedure, as explained in the final topic of this chapter.

3–6 OSCILLATOR DRIFT PROBLEMS

Oscillator drift causes the sound output from a receiver to weaken gradually, unless the tuning dial is reset. That is, the tuning–dial indication and its setting are determined by the operating frequency of the local oscillator. Excessive drift results in impaired receiver sensitivity and interference. The operating frequency of the local oscillator depends not only upon inductance and capacitance values, but also upon device characteristics and bias voltages. As an example, a deteriorating transistor in the oscillator circuit can cause frequency drift. A leaky capacitor, unstable resistor, or moisture are also common causes of drift. Sometimes a marginal defect occurs in an oscillator coil, such as an insulation burn due to a solder splash. Tuner mechanisms become worn after extensive use, and the setting may change as a result of vibration or thermal expansion or contraction. Troubleshooting an oscillator–drift problem accordingly requires a systematic check–out of the various possible causes of instability.

3–7 FM RECEIVER TROUBLESHOOTING

Troubleshooting FM radio receivers involves the same basic principles explained previously for troubleshooting AM radio receivers. However, there are certain technical distinctions to be observed. For example, the signal channel up to the audio–amplifier section processes a frequency-modulated signal. Therefore, signal–injection tests require the use of an FM signal generator. As a practical note in this regard, ordinary service-type AM generators usually have appreciable *incidental FM*, particularly when operated at a high percentage of modulation. In turn, this type of generator can be used for rough signal–injection tests of FM radio receivers.

As exemplified in Fig. 3–11, an FM IF channel operates at 10.7 MHz, instead of 455 kHz, as in the case of an AM receiver. Observe also in Fig. 3–11 that the IF channel includes both 10.7–MHz and 455–kHz IF transformers, inasmuch as this is part of a combination AM/FM receiver. It is possible to operate AM and FM transformers in series, because their resonant frequencies are widely different. Note that the FM AVC function is supplemented by a local–distant switch in the antenna input circuit. This feature minimizes the possibilty of cross–modulation in case unusually strong station signals are present. Diode CR3 is shunted across the oscillator tank circuit to provide automatic frequency control (AFC) and thereby eliminate oscillator drift. However, an AFC defeat switch is also provided so that weak station signals can be tuned in without "skipping over" to a strong adjacent–channel station signal.

Few service shops have specialized FM signal generators available, although the majority of shops employ TV sweep generators that include FM–broadcast and IF frequency coverage. Accordingly, this type of generator is very useful for signal–injection tests in FM receivers, as well as for alignment of the tuned circuits. Observe in Fig. 3–11 that an integrated circuit is utilized in the second AM/FM IF stage. Troubleshooting an IC unit is basically the same as in the case of a transistor unit. That is, the technician checks the stage gain, and proceeds to make DC voltage measurements at the IC terminals, if necessary. An incorrect voltage value indicates a defect either in the IC or in an associated circuit component. Note also that the normal current drain of the IC is specified. In case of doubt, this current drain can be measured. An abnormal or subnormal current reading indicates that a fault is present.

3–8 ALIGNMENT OF RADIO RECEIVERS

AM radio receivers are generally aligned by the peak–response method. That is, alignment adjustments are made with respect to individual specified frequencies supplied by an AM signal generator. On the other hand, wide–band receivers are usually aligned by the sweep–frequency method, as explained subsequently. An AM radio receiver is customarily aligned with a weak test signal, so that the AVC section remains essentially inoperative. The procedure provides a much sharper peak response than if a strong test signal is employed. Unless the receiver service

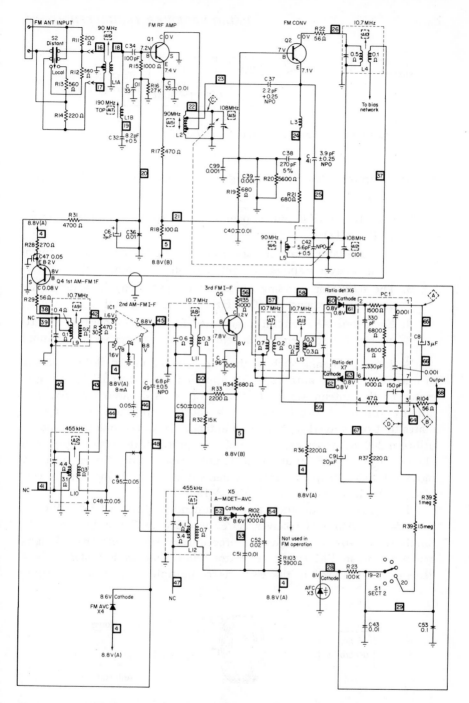

Figure 3–11 RF–IF signal–channel schematic for the FM section of an AM/FM radio receiver (*Courtesy of* Howard W. Sams & Co., Inc.).

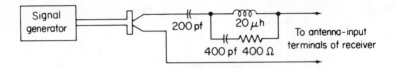

Figure 3–12 Standard dummy antenna for AM radio receiver alignment.

manual instructs otherwise, the alignment procedure follows these con-
secutive steps:

1. Each tuned–IF circuit is peaked at the specified frequency (such
   as 455 kHz).
2. The oscillator trimmer capacitor (or slug) is adjusted for maximum receiver response at 1500 kHz on the tuning dial.
3. The oscillator padding capacitor (if provided) is adjusted for
   maximum response at 600 kHz on the tuning dial.
4. Each tuned–RF circuit is adjusted for peak response at the frequency (or frequencies) specified in the receiver service manual.

A dummy antenna such as the one depicted in Fig. 3–12 is recommended by industry authorities for applying the test signal from a generator to the antenna–input terminals of an AM radio receiver. This
dummy–antenna network provides a source impedance that approxi-
mates that of an average AM radio antenna. However, many technicians
utilize a 200–pF series capacitor only. In practice, the resulting error in
alignment of the RF input circuit is not serious. When the receiver employs a built–in loop antenna, the generator output is inductively coupled to the loop, as shown in Fig. 3–13.

It is instructive to study the recommended alignment procedure for
a solid–state AM/FM receiver, as exemplified in Fig. 3–14. Chart 3–1
tabulates step–by–step alignment procedures, using either an AM signal
generator or an FM signal generator. Note that an FM stereo signal
generator is required in the FM stereo multiplex alignment procedure.

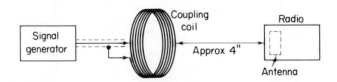

Figure 3–13 A generator coupling coil is used to apply a test signal to a receiver
with a built–in loop antenna.

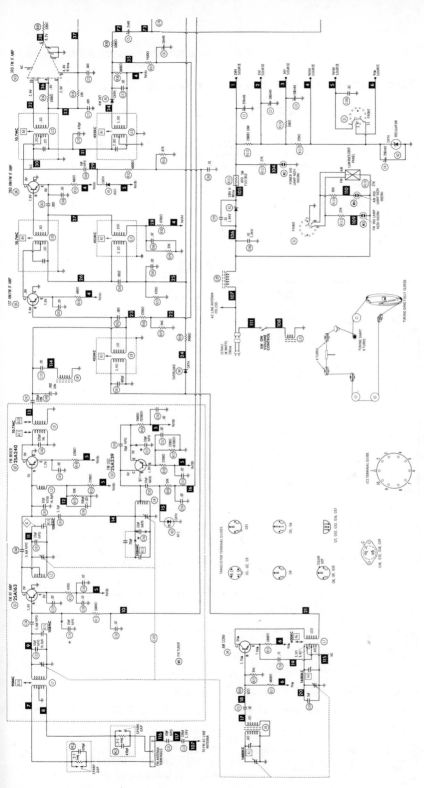

Figure 3–14 Schematic diagram for a combination AM/FM radio receiver (*Courtesy of Howard W. Sams & Co., Inc.*).

Chart 3–1 Alignment procedure (*Courtesy of* Howard W. Sams & Co., Inc.).

Maintain line voltage at 117 volts. Use only enough generator output to obtain a suitable indication.
Allow a 15 minute warmup for receiver and equipment.
CAUTION: Use isolation transformer, if available. If not, observe polarity when connecting test equipment.
Suggested Alignment Tools:
  A1 thru A6, A8, A9, A13, A14, A15...GENERAL CEMENT #8868, 8987, 9089........ WALSCO #2531-X, 2541, 2587
  A7, A10, A11, A12, A16........ GENERAL CEMENT #9296, 9297, 9300....... WALSCO #2510, 2546, 2547
  A17 thru A21.................. GENERAL CEMENT #8606, 8606L, 8869 .....WALSCO #2543, 2544, 2588

AM ALIGNMENT—SELECTOR IN AM POSITION

Fashion loop of several turns of wire and connect generator across loop. Set volume control at maximum.

	GENERATOR FREQUENCY	DIAL SETTING	INDICATOR	ADJUST	REMARKS
1.	455KC (400v Mod.)	Tuning gang fully open.	Output meter across voice coil.	A1, A2, A3	Adjust for maximum. Repeat until no further improvement can be made.
2.	1600KC	"	"	A4	Adjust for maximum.
3.	1400KC	Tune to signal.	"	A5	"
4.	600KC	"	"	A6	Rock tuning gang and adjust for maximum. Repeat steps 2 thru 4 until no further improvement can be made.

FM IF ALIGNMENT USING AM SIGNAL GENERATOR—SELECTOR IN FM POSITION

High side of generator thru .001mfd to point Ⓒ, low side to ground.

	GENERATOR FREQUENCY	DIAL SETTING	INDICATOR	ADJUST	REMARKS
5.	10.7MC (Unmod.)	Point of non-interference.	DC probe of VTVM to point Ⓐ, common to ground.	A7, A8, A9, A10, A11	Adjust for maximum.
6.	"	"	DC probe to point Ⓑ, common to ground.	A12	Adjust for zero reading. A positive or negative reading will be obtained on either side of the correct setting.

FIG. 1

FM IF ALIGNMENT USING FM SIGNAL GENERATOR—SELECTOR IN FM POSITION

High side of generator thru .001mfd to point Ⓒ, low side to ground. Use only enough marker signal for indication. Use 60v frequency modulated signal with 450KC sweep. Use 60v sawtooth voltage in scope for horizontal deflection.

	GENERATOR FREQUENCY	DIAL SETTING	INDICATOR	ADJUST	REMARKS
5.	10.7MC (450KC Swp.)	Point of non-interference	Vert. amp. of scope to point Ⓐ, low side to ground.	A7, A8, A9, A10, A11	Disconnect stabilizing capacitor C8. Adjust for maximum gain and symmetry of response similar to Fig. 1 with marker as shown. Reconnect C8.
6.	"	"	Vert. amp. to point Ⓑ, low side to ground.	A12	Adjust A12 (secondary) to place marker at center of S curve similar to Fig. 2. Adjust A7 (primary) for maximum amplitude and straightness of line.

FM RF ALIGNMENT—SELECTOR IN FM POSITION

Connect generator across antenna terminals with 120Ω carbon resistors in series with each lead.

10.7 MC

	GENERATOR FREQUENCY	DIAL SETTING	INDICATOR	ADJUST	REMARKS
7.	108MC	Set at high end.	DC probe of VTVM to point Ⓐ, common to ground.	A13, A14, A15	Adjust for maximum.
8.	90MC	Tune to signal.	"	A16	Rock tuning and adjust for maximum. Repeat steps 7 and 8 until no further improvement can be made.

FIG. 2

FM STEREO MULTIPLEX ALIGNMENT USING FM STEREO SIGNAL GENERATOR (± .0001% ACCURACY)

High side of generator thru 47K to point Ⓑ, low side to ground.

	GENERATOR FREQUENCY	INDICATOR	ADJUST	REMARKS
9.	67KC	Vert. amp. of scope thru a 47K to point Ⓔ, low side to ground.	A17	Adjust for MINIMUM.
10.	72KC		A18	
11.	19KC	Vert. amp. thru 47K to point Ⓓ, low side to ground.	A19, A20	Adjust for maximum.
12.	"	Vert. amp. thru 47K to point Ⓔ, low side to ground.	A21	Adjust for maximum 38KC response.
13.	Modulated Left Channel	Vert. amp. to point Ⓕ, low side to ground.	A19, A20, A21	Adjust for MINIMUM. This step should require only slight adjustment.
14.	Modulated Right Channel	Vert. amp. to point Ⓖ, low side to ground.		Check for MINIMUM. If necessary make compromise adjustments of A19, A20, A21.

TUNING METER ADJUSTMENT (R5)

With Selector Switch in FM position and tuned off station, adjust R5 until meter needle deflects slightly from bottom of meter.

Multiplex principles are explained in detail in the following chapter. Proficiency in alignment technique is acquired by a combination of study and practical experience.

QUESTIONS

1. What is the first step in troubleshooting principles?
2. What are some of the symptoms of troubles in a receiver?
3. How is trouble such as a weak sound output located?
4. How would you prevent detuning of the resonant circuit when signal–tracing a receiver?
5. What is the purpose of connecting a 0.01 µF capacitor between the generator and the circuit when using the signal–substitution method of signal tracing?
6. How is the local oscillator usually tested for operation?
7. How is a defect located if it is not obvious when the defective stage is located?
8. How are open capacitors best pinpointed if they are in the signal path?
9. Why is an oscilloscope necessary when a signal generator is used for a distortion test?
10. What are some of the causes of distorted output symptoms?
11. What are two sections, other than the speaker, that could be eliminated if the noise level was normal?
12. What are the characteristics of poor selectivity?
13. What are the symptoms of oscillator drift?
14. What are some of the causes of a weak signal?
15. How are AM receivers usually aligned?
16. Why is a weak signal usually used to align an AM receiver?
17. Give four steps for aligning an AM receiver.

4

HI-FI STEREO TROUBLESHOOTING

4–1 GENERAL CONSIDERATIONS

High–fidelity reproduction has not been defined in terms of industry standards. However, it is tacitly agreed that hi–fi operation involves a frequency response that is flat within ±1 dB from 20 Hz to 20 kHz, or more, with a harmonic–distortion value of less than 1%. Stereophonic operation necessitates the use of two separate audio channels, identified as Left and Right (L and R). Accordingly, stereo reproduction employs two audio signals that usually differ in frequency and amplitude. Figure 4–1 shows a typical arrangement of L and R speakers for a hi–fi

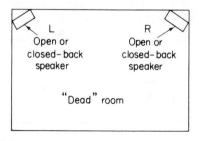

Figure 4–1 Typical arrangement of L and R speakers for a hi–fi stereo system.

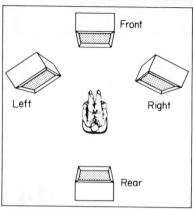

Figure 4–2 Typical arrangement of L, R, Front, and Rear speakers for a 4–channel stereo system.

stereo system. In a four–channel stereo system, four speakers are employed, as exemplified in Fig. 4–2. Four audio signals are utilized, which usually differ in frequency and amplitude.

4–2 CHECKING AUDIO–AMPLIFIER FREQUENCY RESPONSE

An audio amplifier is checked for frequency response by means of the test setup depicted in Fig. 4–3. Note that the speaker is disconnected, and a load resistor is utilized in place of the speaker. This resistor must have a power rating equal to or greater than the maximum rated power output of the amplifier, and a resistance value equal to the rated input impedance of the speaker. An audio TVM is connected across the load resistor to indicate the output voltage. Note that the audio oscillator should have a lower distortion rating than the amplifier under test. The procedure is as follows:

1. Set the audio oscillator to 1 kHz output, and advance the output level until the TVM indicates that the amplifier is producing maximum rated output power. ($P=V^2_{rms}/R_L$.)

2. Set the audio oscillator to 20 Hz, with the same output level as

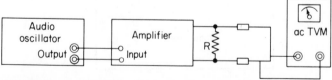

Figure 4–3 Audio amplifier frequency response test setup.

above, and read the dB indication on the TVM scale. Record this value.

3. Repeat the foregoing measurement at progressively higher frequencies, such as 100, 500, 1000, 5000, 10,000, and 20,000 Hz. Plot the measured values as exemplified in Fig. 4–4. Note that the bass and treble controls must be carefully set for flat response.

Poor frequency response can be caused by various component defects, although capacitors are the most likely suspects. For example, with reference to Fig. 4–5, a loss of capacitance in C1 or C2 will result in impaired low–frequency response. The same symptom is produced by a loss of capacitance in C5 or C6. As noted previously, electrolytic capacitors tend to lose capacitance with age. Impaired high–frequency response can result from an "open" in C7 or C8, due to loss of negative feedback. Poor high–frequency response can also be caused by an increase in value of a load resistor, such as R1 through R6. A leaky transistor causes reduced gain and impaired frequency response.

4–3 CHECKING AUDIO–AMPLIFIER HARMONIC DISTORTION

Harmonic distortion measurements are made with the test arrangement depicted in Fig. 4–6. This is basically the same setup as employed for a frequency response check, except that a harmonic distortion meter is also connected across the load–resistor terminals. It is essential that the audio oscillator have a lower distortion rating than the amplifier under

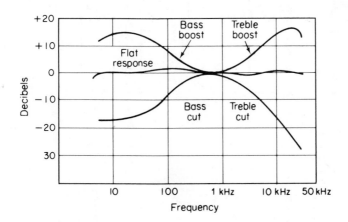

Figure 4–4 Typical frequency response of a hi–fi amplifier, with the variations produced by tone–control settings.

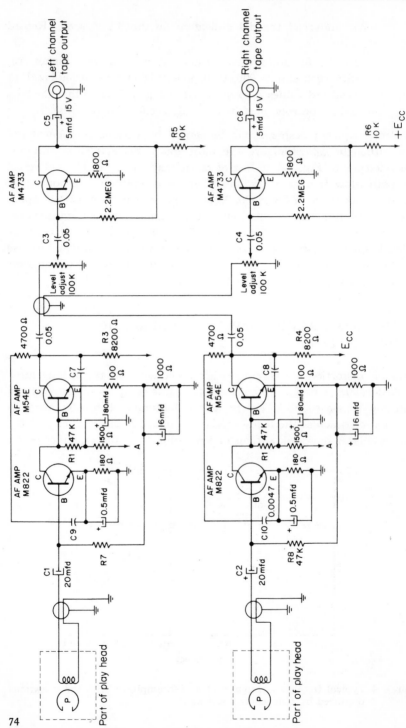

Figure 4–5(a) A typical preamplifier configuration.

test. It is standard practice to measure percentage harmonic distortion at a frequency of 1 kHz, and at maximum rated power output from the amplifier. However, it is good practice to make distortion measurements also at high and low audio frequencies, such as 10 kHz and 50 Hz. An amplifier that has 0.25% distortion at 1 kHz typically produces 0.6% distortion at 10 kHz, and 0.35% distortion at 50 Hz. This frequency dependence of the harmonic–distortion value is attributable chiefly to residual circuit reactances that modify the action of negative–feedback loops.

In general, harmonic distortion tends to increase as the power output from an amplifier is increased. The reason for this increase is seen in Fig. 4–7. Note that the dynamic characteristic of an amplifier is somewhat nonlinear, and that an overload condition occurs at excessive power output. On the other hand, there are a few types of distortion that increase as the power output from an amplifier is decreased. For example, crossover distortion, depicted in Fig. 4–8, produces an increase in the percentage harmonic–distortion reading as the power output from the amplifier is reduced. The reason for this characteristic is that crossover distortion contributes a constant amount of distortion at almost any operating level. At high levels, this distortion component tends to become masked, whereas it dominates the output at low levels. Some forms of hum voltage produce similar results in harmonic–distortion tests.

Excessive harmonic distortion is often caused by an incorrect value of bias voltage, as shown in Fig. 4–9. With reference to Fig. 4–5, bias upset can be caused by leakage in C1, C2, C3, C4, C7, C8, C9, or C10. The defective capacitor can be pinpointed by DC voltage measurements. Bias upset can also be caused by an off–value resistor, such as R7 or R8. Note that if the supply voltage is subnormal, overload distortion will occur at rated power output, due to reduction in the dynamic range of the amplifier. In the case of abnormal ripple in the supply voltage, the resulting hum voltage will raise the indicated percentage of harmonic distortion. Audio technicians often connect an oscilloscope at the output of the harmonic distortion meter, as depicted in Fig. 4–10. The scope is chiefly useful to determine the nature of the distortion products, such as second harmonic, third harmonic, hum, or noise.

4–4 CHECKING AUDIO–AMPLIFIER INTERMODULATION DISTORTION

Intermodulation distortion measurements are made with the test setup shown in Fig. 4–11. A two–tone test signal is employed, which is supplied

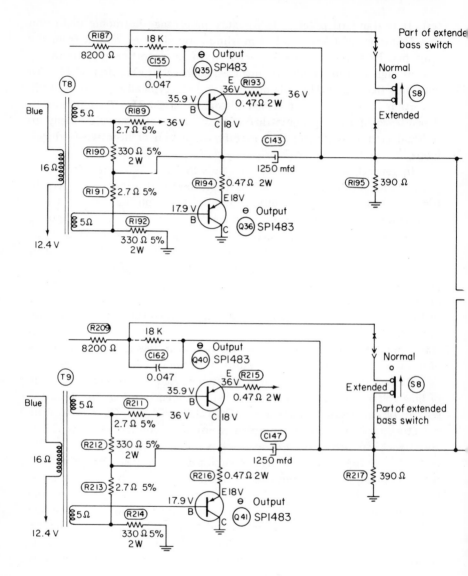

Figure 4–5(b) A typical power–amplifier configuration.

by the intermodulation distortion meter. This signal typically consists of 60–Hz and 6–kHz sine waves. As in the case of a harmonic–distortion test, the amplifier should be checked at maximum rated power output. However, it is good practice to also check the distortion at lower output power levels. The percentage of intermodulation distortion tends to in-

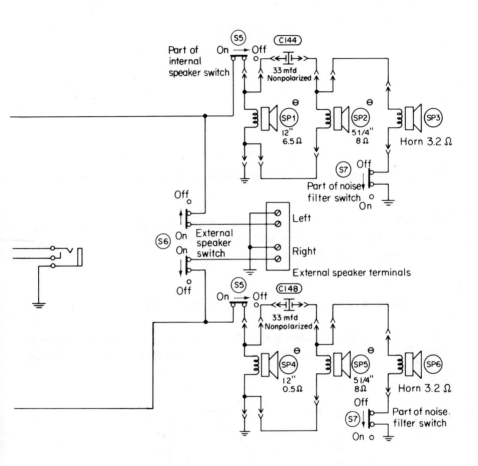

crease as the power output is increased, if the amplifier is operating normally. Figure 4–12 shows a comparison of intermodulation and harmonic distortion for a typical hi–fi amplifier at various power–output levels. Whether a technician utilizes an intermodulation distortion meter or a harmonic distortion meter is chiefly a matter of personal preference.

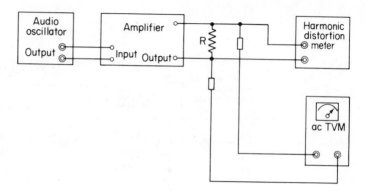

Figure 4–6 Test setup for a percentage harmonic–distortion determination.

Note that the same amplifier faults that cause an abnormally high harmonic–distortion reading will also cause an abnormally high inter-modulation–distortion reading.

4–5 STEREO–MULTIPLEX TROUBLESHOOTING

As noted previously, a stereo signal energizes two separate audio chan-nels. The configuration in Fig. 4–5 typifies stereo channels in an audio

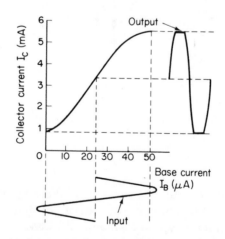

Figure 4–7 The dynamic characteristic of an amplifier is nonlinear, becoming most pronounced in overloaded operation.

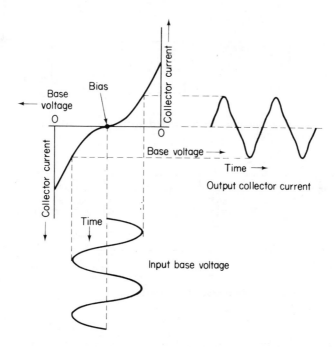

Figure 4–8 Crossover distortion produces a higher percentage of harmonic distortion at low output levels than at high output levels.

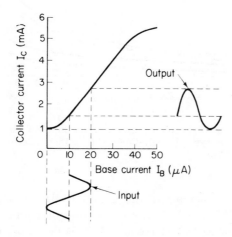

Figure 4–9 Harmonic distortion is often caused by an incorrect value of bias voltage.

79

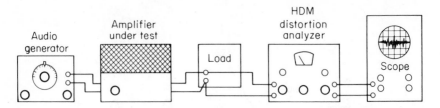

Figure 4–10 Oscilloscope may be connected at the output of a harmonic distortion meter.

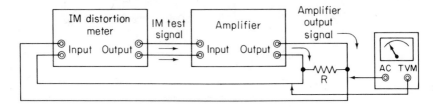

Figure 4–11 Intermodulation distortion test setup.

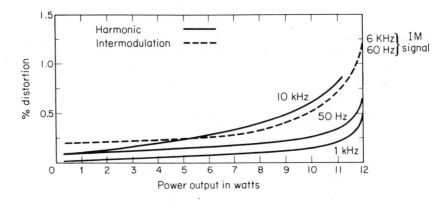

Figure 4–12 Comparison of harmonic and intermodulation distortion percentages for a typical high–fidelity amplifier (*Courtesy of* General Electric Co.).

system. In FM–stereo operation, the L and R signals are multiplexed (encoded) in the FM signal. Therefore, a multiplex unit is required to decode the demodulated FM–stereo signal. This decoding process results in separation of the L and R audio signals from the composite stereo signal. In turn, the separate L and R signals are fed to the L and R channels of an audio amplifier. A block diagram for an FM–stereo receiver is shown in Fig. 4–13. Observe that the multiplex unit is comprised of a bandpass amplifier, 38–kHz oscillator, detector, and matrix in this example.

It is instructive to consider the waveforms that are involved in a stereo–multiplex signal. With reference to Fig. 4–1, Left (L) and Right (R) speakers are utilized at the receiver. These speakers correspond to L and R microphones at the FM transmitter, as depicted in Fig. 4–14. In this example, the outputs from the L and R microphones are being combined or added together to form an L+R audio signal. Note carefully that this is essentially the same signal that would be produced by operation of a single microphone in an ordinary FM transmitter. That is, the arrangement in Fig. 4–14 produces a monophonic FM signal. This mono signal has the basic waveform shown in Fig. 4–15.

Next, let us observe how a mono signal is multiplexed into a stereo signal. First, a 38–kHz subcarrier is inserted with the mono or L+R signal, as depicted in Fig. 4–16. Since the subcarrier frequency is above the range of audibility, it does not affect transmission and reception of the L+R audio signal. Next, the 38–kHz oscillator is amplitude–modulated by another audio signal, called the L—R signal, as shown in Fig. 4–17. This L—R audio signal also occupies a frequency range from zero to 15 kHz, but since it is modulated on the subcarrier, it extends from 23 kHz in the signal spectrum. That is, the L—R signal is above the range of audibility and does not affect transmission and reception of the L+R signal.

Now, it is important to observe the formation of L+R and L—R signals from L and R microphones at the transmitter, because the multiplex unit in the receiver simply operates "backward" in decoding the composite stereo signal. Figure 4–18 shows how the outputs from the L and R microphones are combined in a mixer to form the L+R signal. Note that the output from the R microphone is also passed through a phase inverter to form a —R signal. Then, this —R signal is mixed with the L signal to form the L—R signal. Thereby, the L and R signals are processed for encoding into the composite stereo signal. At the receiver, a multiplex unit will decode this waveform to recover the original L and R audio signals.

When the L+R and L—R signals are formed as shown in Fig. 4–18, the L—R signal is utilized to amplitude–modulate the subcarrier oscillator, and the resulting L—R modulated wave is mixed with the L+R signal, after which this composite stereo signal is frequency–modulated on an RF carrier for transmission, as depicted in Fig. 4–19. The composite stereo signal has the waveform exemplified in Fig. 4–19(b). Ob-

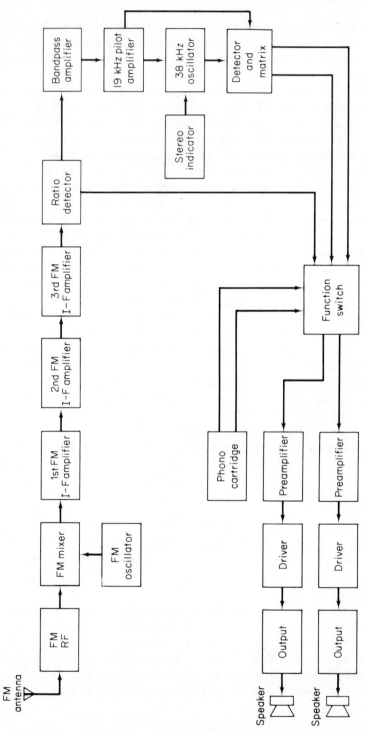

Figure 4-13 Plan of an **FM**-stereo radio receiver.

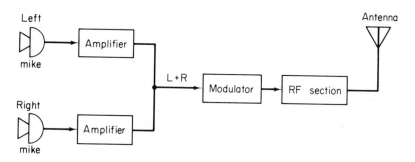

Figure 4–14 Simplest arrangement of L and R microphones at an FM transmitter.

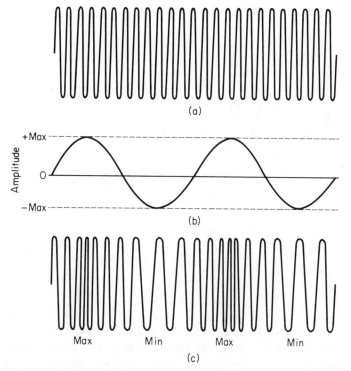

Figure 4–15 Basic monophonic FM signal waveform. (a) Carrier wave; (b) Modulating audio signal; (c) Frequency–modulated carrier wave.

serve carefully that this composite stereo signal has the R audio signal encoded as the upper modulation envelope, and has the L audio signal encoded as the lower modulation envelope. Some FM–stereo receivers employ multiplex units that have envelope detectors to decode the composite stereo signal.

83

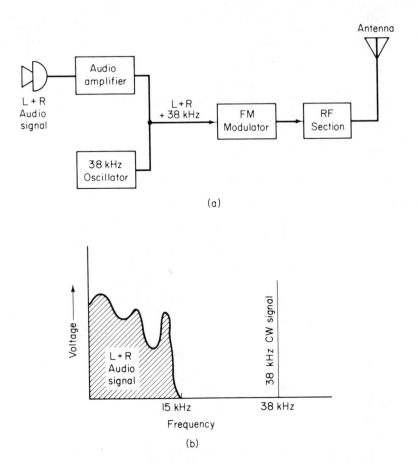

Figure 4–16 Insertion of a 38–kHz subcarrier. (a) Block diagram; (b) Frequency spectrum.

Note that the arrangement shown in Fig. 4–19(a) has been simplified for instructive purposes. In actual practice, a balanced modulator is employed to modulate the L—R signal on the 38–kHz subcarrier, in order to suppress the subcarrier and retain only the sidebands. Subcarrier suppression is utilized in order to reduce the energy content of the composite stereo signal, so that the optimum signal–to–noise ratio is realized. In turn, of course, the suppressed 38–kHz subcarrier must be reinserted at the stereo–FM receiver, in order to reconstitute the composite stereo signal. This is a subfunction of the multiplex unit at the receiver. Since the 38–kHz subcarrier must be regenerated in precise frequency and phase by the multiplex unit, synchronization is required.

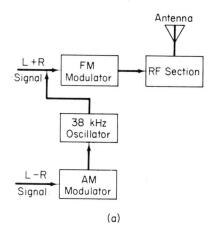

(a)

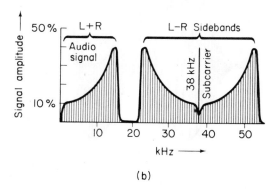

(b)

Figure 4–17 Modulation of the subcarrier by an L–R audio signal. (a) Block diagram; (b) Frequency spectrum.

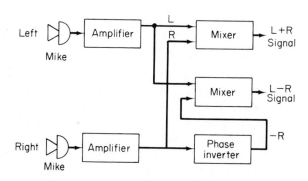

Figure 4–18 Formation of the L+R and L—R signals from the L and R signals.

85

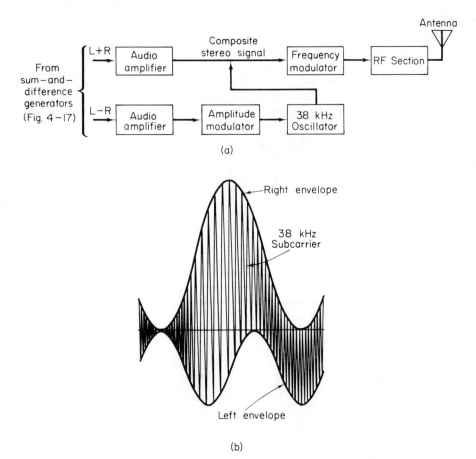

Figure 4–19 Modulation of the FM transmitter by the composite stereo signal. (a) Block diagram; (b) Composite stereo waveform.

This is provided by means of a low–level 19–kHz pilot subcarrier, as depicted in Fig. 4–20.

Three chief stereo–multiplex decoding arrangements are used in FM receivers, as depicted in Fig. 4–21. In the bandpass–and–matrix arrangement, the L+R and L—R signals are added and subtracted in mixers to form the L and R signals. Note that the L—R sidebands are separated from the L+R signal by means of a 23– to 53–kHz bandpass circuit. Next, the output from the bandpass circuit is mixed with a 38–kHz frequency to reconstitute the L—R signal. An AM detector recovers the L—R audio envelope from the reconstituted L—R signal. Finally, the L—R signal is added to the L+R signal to obtain a 2L signal. Also,

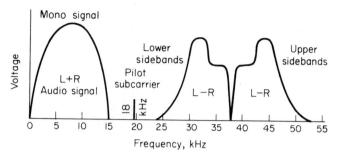

Figure 4–20 Spectrum of the stereo–FM signal, as it arrives at the receiver.

the L—R signal is subtracted via a phase inverter from the L+R signal to obtain a 2R signal. Note that the 38–kHz oscillator must be synchronized by a 19–kHz pilot–subcarrier processing section (not shown in the diagram). A standard frequency–response curve for the bandpass circuit is shown in Fig. 4–22.

In the electronic–switching method depicted in Fig. 4–21(b), the regenerated 38–kHz subcarrier alternately samples the positive and negative envelopes of the composite stereo waveform. As seen in Fig. 4–19(b), these samples follow the L and R waveforms, whereby the composite stereo waveform is decoded. Finally, in the envelope–detection method depicted in Fig. 4–21(c), the reconstituted composite stereo waveform is applied to a positive–rectifying detector and to a negative–rectifying detector. In turn, the composite stereo waveform is decoded into its constituent L and R signals. In normal operation, each of these decoding methods is capable of providing hi–fi reproduction with good separation of the L and R audio signals.

Figure 4–23 shows the configuration for an electronic–switching multiplex unit. The composite audio input comprises the L+R signal mixed with the L—R sidebands and the 19–kHz pilot subcarrier. This input signal branches into a 19–kHz amplifier circuit and into the switching bridge consisting of diodes CR3, CR4, CR5, and CR6. A frequency doubler (CR1 and CR2) follows the 19–kHz amplifier, in order to regenerate the 38–kHz subcarrier. In turn, the subcarrier signal is mixed with the incoming audio signal to reconstitute the composite stereo waveform. Bridge conduction is determined by the instantaneous polarity of the subcarrier driving voltage, and the two sides of the bridge conduct alternately. Accordingly, the positive and negative envelopes of the composite stereo waveform are sampled alternatively at a 38–kHz rate, and the bridge outputs consist of the decoded L and R signals.

The most basic test of a multiplex unit concerns the number of dB

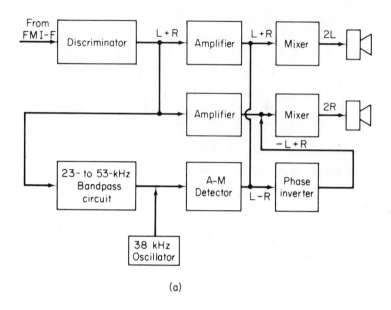

(a)

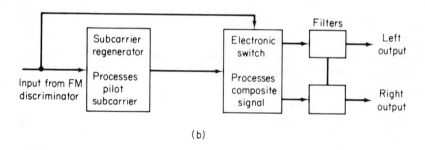

(b)

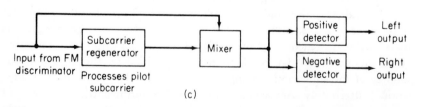

(c)

Figure 4–21 Basic types of stereo–multiplex decoders. (a) Bandpass–and–matrix
arrangement; (b) Electronic switching method; (c) Envelope detection method.

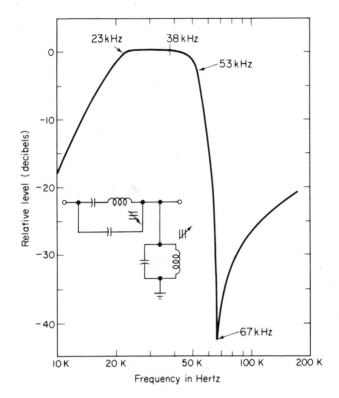

Figure 4–22 Standard frequency–response curve for the bandpass filter circuit in a stereo–multiplex unit.

separation that it provides between the L and R signals. Figure 4–24 shows the arrangement for a stereo multiplex separation test. L and R signals are applied in turn from a stereo multiplex signal generator. A composite audio signal is employed in this procedure, as illustrated in Fig. 4–24(b). Note that the waveform appears to be the same, whether the generator is supplying L output or R output. The reason for this similarity is that L and R composite audio signals differ only in phase. When an L signal is applied in the separation test, there is theoretically zero output from the R channel, and maximum output from the L channel. Similarly, when an R signal is applied, there is theoretically zero output from the L channel, and maximum output from the R channel. In practice, however, about 30 dB difference will be observed between the TVM readings, for a normally operating stereo multiplex unit. A separation of 10 dB is considered to be on the borderline of a trouble symptom.

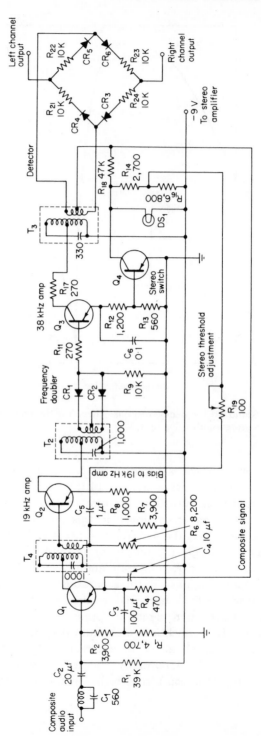

Figure 4–23 Configuration of an electronic–switching multiplex unit.

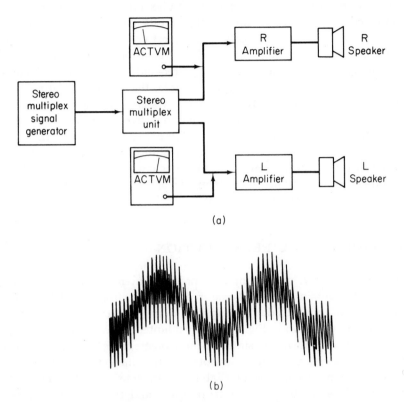

Figure 4–24 Stereo multiplex separation test. (a) Test setup; (b) Waveform supplied by stereo multiplex signal generator.

Inadequate separation is usually caused by a component defect in the multiplex unit, such as an "open" or "shorted" capacitor. Note, however, that misalignment can also cause poor separation. This is particularly important in a matrix–type unit that operates with the frequency response curve depicted in Fig. 4–22. Multiplex alignment procedure is specified in the pertinent service manual. Other common causes of poor separation are unmatched diodes, or diodes with subnormal front–to–back ratios in a switching bridge, leaky capacitors in a switching bridge, off–value resistor in a matrix circuit, or collector leakage in a transistor. Sometimes it is found that a stereo multiplex unit has normal separation when tested with a composite audio signal, as shown in Fig. 4–24, but exhibits poor separation when energized through the FM receiver. This discrepancy points to poor alignment of the high–frequency signal channel in the receiver.

Distorted output from a stereo multiplex unit can be caused by a defective capacitor in a switching bridge, faulty electrolytic capacitor in the subcarrier regeneration section, a defective diode in an envelope detector or switching bridge, or a failing transistor in an amplifier section. Finally, it should be noted that stereo–indicator failure is a comparatively common complaint. For example, the stereo–indicator lamp DS1 in Fig. 4–23 may fail to glow when a stereo signal is present. In case the stereo threshold adjustment has no effect, the indicator lamp is probably burned out. If further troubleshooting is required, transistor Q4 should be checked. Malfunction also results if C6 is seriously leaky, or if a resistor in the indicator circuit has changed value substantially. DC voltage and resistance measurements are the most useful clues in pinpointing the defective component.

4–6 QUADRAPHONIC REPRODUCTION

As noted previously, quadraphonic reproduction entails the use of four speakers. To obtain quadraphonic reproduction from an FM stereo signal, a quadraphonic adapter and two extra speakers are added to a stereo receiver as shown in Fig. 4–25. The adapter processes the stereo signal in phase and frequency to simulate the acoustic reflections typically occurring in theaters, nightclubs, and other entertainment sites. Note that true four–channel sound means that there are four discrete (different) sound sources at both the program terminal and the reproduction terminal. True quadraphonic sound therefore requires true four–channel program material, and, at the reproduction terminal, four channels of reproduction (four amplifiers and four speakers).

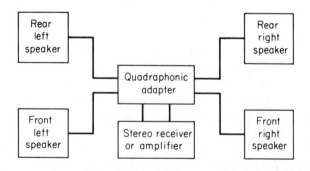

Figure 4–25 Quadraphonic adapter arrangement for obtaining synthesized four–channel sound.

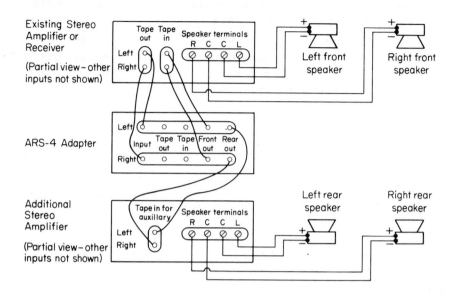

Figure 4–26 Typical quadraphonic adapter connections.

Next, synthesized four–channel sound denotes a method whereby conventional two–channel stereo material is reprocessed to make each of four speakers energized by somewhat different signals, thereby simulating a true four–channel effect. Synthesized four–channel sound is not true four–channel sound. A quadraphonic adapter is connected as exemplified in Fig. 4–26, and passes a conventional stereo signal through RC networks, thereby providing synthesized four–channel sound. Troubleshooting a quadraphonic adapter is a simple procedure that generally requires only capacitor tests. That is, an "open" or "shorted" capacitor in the adapter circuitry will weaken and distort the sound output.

QUESTIONS

1. What are the requirements of the resistor that is used to replace the speaker when aligning an audio amplifier?
2. Why does harmonic distortion usually increase as the output power increases?
3. What are some of the causes of excessive harmonic distortion?

4. Why is it a personal preference whether an intermodulation distortion meter or a harmonic distortion meter is used to test an amplifier?

5. What are three types of stereo–multiplex decoding arrangements?

6. If it is found that a stereo system has normal separation under a composite audio signal test, but poor separation when connected to an FM receiver, what is the trouble?

5

TAPE RECORDER
TROUBLESHOOTING

5-1 RECORDER HEADS

A tape-recording system is arranged as shown in Fig. 5–1, and the play-back system is arranged as shown in Fig. 5–2. The recording head depicted in Fig. 5–1 is basically an electromagnet energized from an audio amplifier. In turn, the electromagnet produces a magnetic field that varies in strength at an audio–frequency rate. This varying magnetic field is impressed on a moving tape which is coated with a ferromagnetic

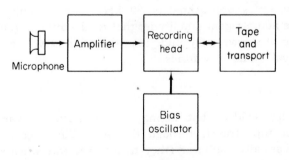

Figure 5–1 Plan of a tape–recording system.

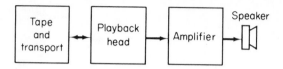

Figure 5–2 Plan of a tape–player system.

substance. The result is that the recording tape becomes permanently magnetized in accordance with the magnetic field variation, and the audio–frequency current that is energized with the recording head in Fig. 5–1 can be reproduced by the system depicted in Fig. 5–2.

Recording heads have a highly restricted magnetic field, as seen in Fig. 5–3. Economy–type tape recorders use the same head both for recording and for playback. However, a playback head should have a gap that is sufficiently narrow to provide good fidelity without running the tape at excessively high speed. Therefore, deluxe tape recorders employ separate recording and playback heads. It is advantageous to utilize a somewhat wider gap in a recording head, to impress high-intensity magnetic patterns on the tape. A dual–purpose head with a compromise gap width incurs some loss of quality in reproduction, although it is acceptable to listeners other than high–fidelity buffs. Note in Fig. 5–3 that two gaps are provided in each of the heads. This arrangement is used for stereophonic recording on two tracks simultaneously.

If a tape is sprinkled with carbonyl iron powder after it has passed through a recording head, the magnetic field that was impressed on the ferromagnetic coating becomes visible, as exemplified in Fig. 5–4. To anticipate subsequent discussion, the prominent vertical "valleys" in Fig. 5–4 are the zero points of the AC bias voltage (see bias oscillator block in Fig. 5–1). That is, these "valleys" are not zero points in the audio waveform, which has a maximum frequency of about 15 kHz. The bias-oscillator frequency is approximately 50 kHz, and its amplitude is greater than the audio–signal amplitude; that is, the audio signal varies more slowly and at a lower amplitude than the AC bias voltage. Figure 5–5 shows typical audio waveforms.

5–2 AC BIAS

From Fig. 5–4, it is evident that the magnetic field strength varies considerably as the tape travels through the head. Since ferromagnetic substances are basically nonlinear (they tend to saturate when strongly magnetized), it is necessary to provide some means of linearization. This

Figure 5–3 Recording heads. (*Courtesy of* Howard W. Sams & Co., Inc.).

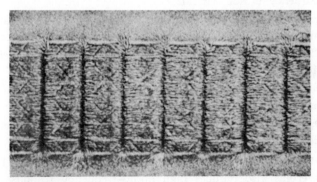

Figure 5–4 Magnetic dust shows pattern impressed on a recorder tape (*Courtesy of* 3–M Co.).

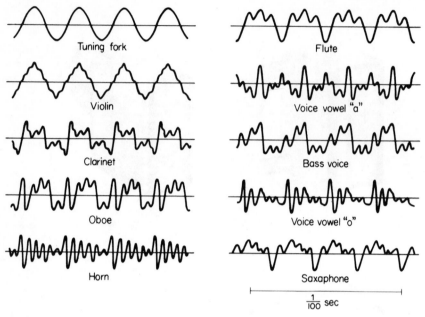

Figure 5–5 Typical audio waveforms.

is accomplished by means of an AC bias voltage, noted previously. When this ultrasonic voltage is mixed with the audio–frequency voltage, as depicted in Fig. 5–1, the ferromagnetic coating of the tape becomes effectively linearized over a useful range of magnetic field strengths. This AC bias tone is inaudible on playback, and only the linearized audio signal is heard. Troubleshooting the AC bias system requires recognition of the following factors:

1. There is an optimum amplitude of bias voltage, although its frequency is not critical.
2. In addition to linearization, the bias voltage also affects the audio–frequency response that can be realized in the recording.
3. The bias voltage also affects the output level of the recording on playback.
4. Peak bias is defined as the value of bias voltage that provides maximum output level on playback.
5. Optimum fidelity or linearization (minimum distortion) is provided by overbias (bias greater than peak) of 2 dB.
6. Optimum low audio–frequency response is obtained at peak bias.
7. Overbias also reduces high audio–frequency response, but improves fidelity, as noted above.

8. Underbias causes distortion, poor signal–to–noise ratio, and reduced output. Bias adjustments are generally provided in tape recorders, as seen in **Fig. 5–6**.

Observe also in Fig. 5–6 that an erase head is depicted. Erasure of a magnetic tape removes the previous recording so that it is returned to an unmagnetized state. Erasing is accomplished by means of a head with a wide gap, which is energized by a high–level bias voltage. Note in Fig. 5–6 that the level of the erase voltage is not adjustable.

Heads in reel–to–reel tape recorders can readily be cleaned manu-

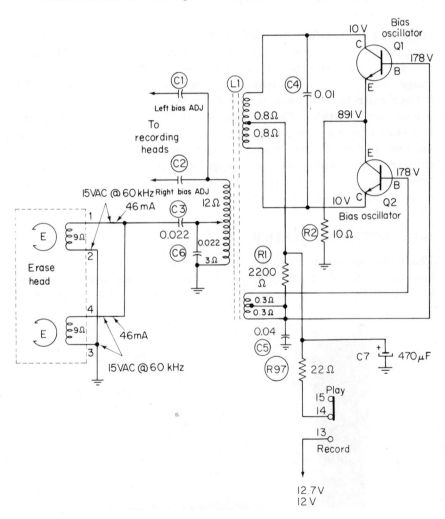

Figure 5–6 Typical bias–oscillator arrangement.

Figure 5–7 A tape lube kit (left), and a clean–lube kit (right) (*Courtesy of* Robins Industries Corp.).

ally, using a clean–lube kit such as the one depicted in **Fig. 5–7.** However, in the case of cassette or cartridge recorders or tape decks, it is necessary to utilize a tape–head and capstan–cleaning cassette or cartridge, such as the one illustrated in **Fig. 5–8.** This is a unit that fits 8–track cartridge machines. It is inserted into the machine instead of a conventional cartridge. The machine is turned on, and the cleaning tape is operated for a specified length of time. Note that each time the cleaning tape makes one revolution, a beeping sound is reproduced by the speaker. The cleaning tape is stopped and removed after three beeps have sounded. It is recommended that 8–track heads be cleaned after each 40 hours of use.

5–3 ELECTRONIC SYSTEM

Economy–type recorders utilize the same amplifier system both for recording and for playback. Deluxe designs employ separate recording and playback amplifiers, as depicted in **Fig. 5–9.** Observe the configuration

Figure 5–8 An 8–track tape–head cleaning cartridge.

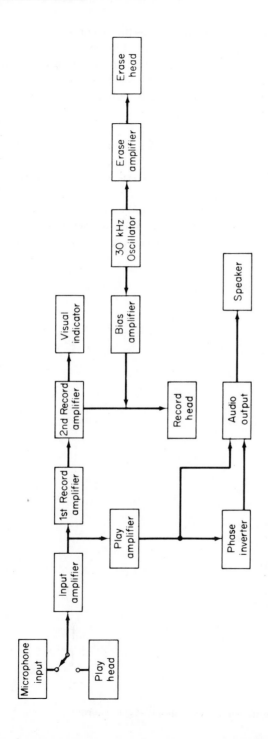

Figure 5–9 A tape–recorder arrangement that employs separate recording and playback amplifiers.

for an economy–type recorder in **Fig. 5–10.** A compromise record/play head is used, with a separate erase head. The same amplifier is utilized both for recording and for playback. A microphone provides an output of several millivolts, and a typical playback head provides an output of 8 millivolts. Therefore, a comparatively high–gain audio–amplifier system is required. In stereophonic recorders, two separate audio channels are provided.

Malfunctions can result in trouble symptoms such as weak or no recording function, poor high–frequency response, poor low–frequency response, distortion, noise, hum, or interference. For example, microphones sometimes become defective. A microphone can be checked with a sensitive **TVM** (audio voltmeter) or with a sensitive oscilloscope. Otherwise, a substitution test is advisable. Sometimes, poor high–frequency response is caused by the use of an excessively long input cable.

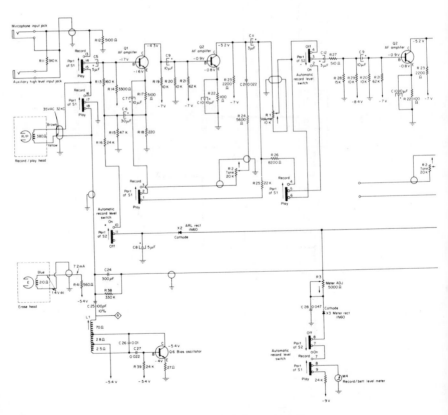

Figure 5–10 Configuration of an economy–type recorder.

That is, a microphone is necessarily used with a shielded cable, and the cable shunts effective capacitance across the microphone circuit. Excessive bypassing action results in impaired high–frequency response.

Some microphones have higher output than others. Therefore, a recorder designed for use with a high–level microphone will not operate properly with a low–level microphone. Weak recording action can also be caused by subnormal bias voltage. The bias voltage can be measured with a TVM. If the trouble is localized to the bias oscillator, check first for a leaky or open capacitor. If the transistor is defective, its DC terminal voltages will usually be incorrect. A recording head may become defective; a substantial number of shorted turns, or an open circuit, can be checked with an ohmmeter. For example, the R/P head in Fig. 5–10 has a rated resistance of 380 ohms.

Heads are also subject to wear, and will produce weak and distorted

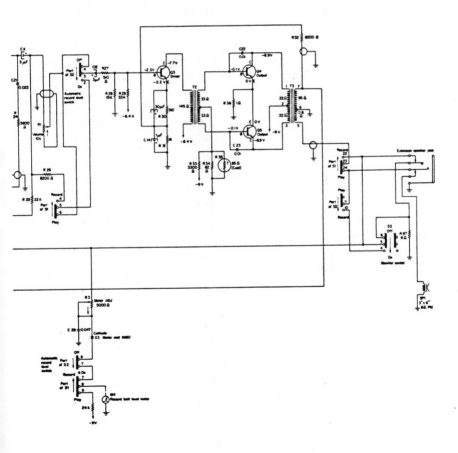

recordings if excessively worn. A replacement head should match the original, and must also be mounted precisely at a right angle to the line of tape travel. Malfunctions in amplifiers can also result in weak or no recording action. Amplifiers are analyzed by means of the same trouble-shooting procedures that were explained in Chapter 4. Sometimes an elementary error is responsible for difficulties in recording, such as threading a tape through the head assembly with the ferromagnetic coating on the opposite side from the gap. Or an accumulation of dirt or foreign substance on the head may be preventing the ferromagnetic coating from contacting the pole pieces.

Poor high–frequency response can be caused by a reversed tape, excessively 'iigh bias voltage, incorrect setting of the tone control, a defective microphone, or a magnetized recording head. It is good practice to demagnetize the head at intervals with a standard demagnetizer (Fig. 5–11). An amplifier defect may also cause poor high–frequency response, as explained previously in Chapter 4. Poor low–frequency response may be caused by an incorrect setting of the tone control, use of an incorrect type of microphone, or by an amplifier defect. Overbias impairs low–frequency response as well as high–frequency response. Bias voltage can be measured with a TVM, and compared with the value specified in the recorder service data.

Distortion can be caused not only by an incorrect value of bias voltage, but also by a distorted bias–voltage waveform (that is, the bias voltage should have a good sine waveform). This characteristic can be checked with an oscilloscope. In the case of a distorted bias waveform, look for a defect in the bias–oscillator circuit. For example, an incorrect value of replacement capacitor can provide a normal output level from the oscillator, but with a highly distorted waveform. A common cause of distortion in the case of an inexperienced operator is use of an excessively high recording level. If the recording level appears to be normal, and distortion persists, check the recording–level meter circuit. Amplifier defects may cause distortion, as discussed in Chapter 4.

Figure 5–11 A tape–head demagnetizer (*Courtesy of* Robins Industries Corp.).

Noise, hum, or interference in recordings can be caused by worn and defective tape. For example, a tape that has been erased and reused many times will eventually become noisy. Defective microphones can also cause noise. Hum is usually the result of a poor ground connection, such as that caused by a frayed cable. Noise can also be caused by defective erase action. In the example of Fig. 5–9, AC erase is employed, and the erase voltage can be monitored with an oscilloscope to determine whether it may be erratic. Again, in the example of Fig. 5–10, DC erase is utilized (that is, direct current is passed through the erase head). The DC erase voltage can be measured and monitored with a TVM. Interference is commonly caused by operation in strong fields, such as in the proximity of a radio transmitter. In such a case, the recorder should be relocated. Finally, noise, hum, or interference can be caused by amplifier defects, as explained in Chapter 4.

It will be found that different types of tape require different bias levels. Small monophonic tape recorders generally utilize iron–ferrite tapes, and employ comparatively low–amplitude bias. On the other hand, elaborate machines may also use chromium–dioxide tapes that require comparatively high–amplitude bias. In turn, poor results will be obtained if an attempt is made to record on chromium–dioxide tape at a low bias level. Therefore, the technician must keep in mind that special bias requirements are a necessity in such a case.

5–4 BASIC MECHANICAL DIFFICULTIES

A tape–drive assembly is employed to move the tape past the heads at a constant speed while the recorder is in operation. A rewind function is also provided in many machines which permits the tape to be reeled up rapidly. Figure 5–12 depicts the arrangement of a tape cassette. While the tape is being recorded or played back, it is pressed against a rotating shaft (capstan) by a pinch roller. The capstan function is provided by a flywheel, and is driven by an electric motor. A speed–reduction arrangement is provided between the motor and capstan. This ordinarily consists of a belt–and–pulley assembly.

Defects in the mechanical system can cause operating symptoms such as no movement of the tape, incorrect tape speed, faulty braking action, erratic operation, and tendency of the tape to ride up and down between the capstan and the pressure roller. Troubleshooting should start with a tape that is in good condition. A tape that has been spliced may "foul up" between the capstan and the pressure roller. Next, the capstan and pressure roller should be inspected, and cleaned if necessary. Figure 5–13

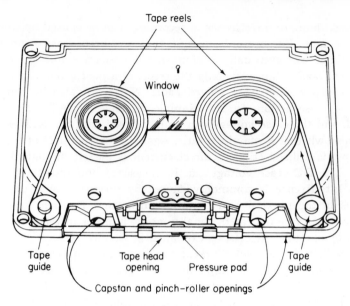

Figure 5–12 Arrangement of a tape cassette.

Figure 5–13 Appearance of a capstan and flywheel, pressure roller, and spring (*Courtesy of* Howard W. Sams & Co., Inc.).

shows the appearance of a capstan and flywheel, pressure roller, and spring. If the spring is weak or broken, tape–transport trouble will result. Worn or fouled belts also cause failure of the tape to move uniformly. A belt is replaced as shown in Fig. 5–14.

Incorrect tape speed or erratic transport can be caused by a dragging brake shoe. Figure 5–15 shows the appearance of a typical brake and shoe. The mechanism should be adjusted, if necessary, to eliminate any drag during forward transport. Do not overlook the possibility of lack of lubrication, or fouled bearings that impose abnormal friction. Faulty braking action is commonly caused by a weak or broken brake spring, which must then be replaced with a spring that provides correct tension. Loose screws can cause a brake shoe to fail to retract. Sometimes lubricating oil crawls over the braking surface and causes braking fail-

Figure 5–14 Belt replacement in a tape recorder (*Courtesy of* Howard W. Sams & Co., Inc.).

Figure 5–15 A brake and shoe mechanism (*Courtesy of* Howard W. Sams & Co., Inc.).

ure. Gummy deposits will cause the brake to grab and break the tape. If brake pads are utilized, the pads must be replaced when they are worn out.

When the tape tends to ride up and down between the capstan and the pressure roller, there is a possibility of lubricating oil on the capstan surface. Check the pressure–roller mounting stud, to make certain that it is parallel to the capstan, and replace with a new pressure–roller assembly, if necessary. Again, the capstan surface may be scratched or scored, and requires replacement. Excessive take–up torque is another possibility—inspect the clutch or the belt assembly for abnormalities. Sometimes a pressure roller becomes eccentric or flattened and causes the tape to wander up and down.

Figure 5–16 shows track placement for two, four, and eight track recording. Note how each track must be aligned precisely with its corresponding gap in the head. Typical head–adjustment facilities are illustrated in Fig. 5–17. The height is adjusted to match the track on a standard test tape, as indicated by maximum output. The azimuth adjustment refers to the angle made by the gap to the line of tape travel, and should be exactly 90°. Maximum output and minimum distortion is obtained when this azimuth adjustment is correct. It is desirable to use a standard test tape, because care is taken to record the test tone from a precisely set gap. In other words, if a tape is employed that has been recorded from a head with an incorrect azimuth setting, it will play back satisfactorily through the same head. On the other hand, a pre–recorded tape from a correctly adjusted head will not play back properly through the head with incorrect azimuth setting.

Tape speed is checked to best advantage with a stroboscope disc, viewed under a neon or fluorescent lamp. If the tape speed is incorrect, off–pitch reproduction will result. Wow or flutter will show up as a tendency of the strobe bars to vary in speed back and forth, or to start rotating in one direction and then to stand still momentarily and then start rotating in the other direction. Special strobe tapes are also available to make speed checks. In the case of incorrect tape speed, inspect the governor (if used), belts, and bearings. If the contacts on a governor are defective, the motor will speed up excessively. Gummed bearings or loose belts tend to make the tape run too slow. Motors cause comparatively little trouble, and if a motor becomes defective, it is generally most expedient to replace it.

Many 8–track machines are used in automobiles and operate from a 12–volt battery. Therefore, the technician needs a good 12–volt battery eliminator on his bench. As noted previously, a solenoid is generally used to change tracks. This solenoid will draw a surge as high as 10

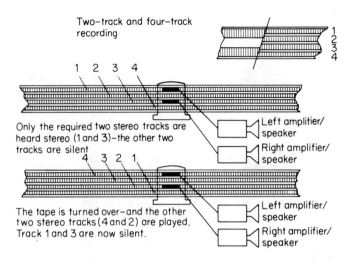

Two–track and four–track recording

Only the required two stereo tracks are heard stereo (1 and 3)–the other two tracks are silent

Left amplifier/ speaker

Right amplifier/ speaker

The tape is turned over–and the other two stereo tracks (4 and 2) are played. Track 1 and 3 are now silent.

Left amplifier/ speaker

Right amplifier/ speaker

Eight–track recording

5–16 Two, four, and eight track recording.

amperes. In turn, the bench power supply must be capable of providing this current demand. Various 8–track test tape cartridges are also needed for checking speed, azimuth, wow, crosstalk, and flutter. It is also good practice to keep some empty cartridges available, in case a machine breaks the tape when operated upside down. This is a result of abnormal tape drag, and can be corrected in most cases by replacing the cartridge. Note that processed "dolbyized" tapes should never be used for test work, because they might give false clues regarding troubles to the un-

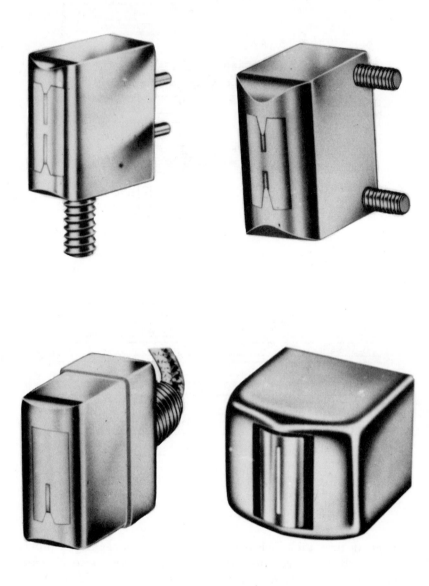

Figure 5–17 Typical head adjustments provided on various tape recorders. (a) Typical bottom–mounted tape head; (b) Typical side–mounted head assembly; (c) Example of rear–mounted playback head; (d) A "no–mount" record/playback head in its bracket (*Courtesy of* Howard W. Sams & Co., Inc.).

initiated. The Dolby system provides minimum noise reproduction, and, because of its unconventional design and relative complexity, requires specialized attention.

Typical trouble symptoms for 8–track players are failure to change tracks, complaint of slow transport or wow, or an annoying rattle which might be produced when the car goes over a bump in the road. Sometimes all three trouble symptoms might be present. With reference to Fig. 5–18, the head carriage must be under adequate tension; otherwise the track change will not occur although the solenoid operates normally, or the head carriage might be binding for some reason such as foreign matter in the mechanism. The tension spring can be lifted a trifle to determine whether or not the head carriage is binding. If the carriage is free, the tension spring should be bent slightly to increase the tension.

Slow transport and/or wow can be caused by defective motor bearings or by foreign matter in the capstan assembly. Another common offender is a worn and stretched drive belt. Note the punch arm indicated in Fig. 5–18. It must seat properly in the locking slot at the side of the cartridge. It is mounted on a post that can become bent. If this occurs, the pinch roller inside the cartridge is prevented from pressing against the capstan shaft with adequate force. Rattling when the machine is vibrated or jarred is generally due to a loose fit between the solenoid plunger and the guide for the plunger. The best procedure is to replace the plunger assembly. However, if it is difficult to procure replacement parts, thin pieces of insulating tubing may be slid over the "wings" of the guide, so that the plunger does not rattle in its mounting.

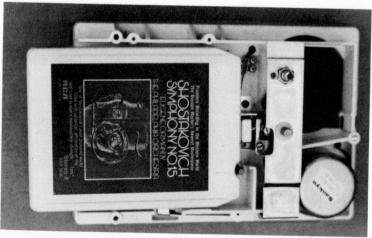

Figure 5–18 Arrangement of a typical 8–track player (*Courtesy of* Howard W. Sams & Co., Inc.).

QUESTIONS

1. Basically, what is the recording head of a tape recorder?
2. How are audio signals placed on a magnetic tape?
3. How is the nonlinear characteristic of a magnetic tape overcome?
4. What is the typical output of the recording head?
5. What is the purpose of the erasure head in a tape recorder?
6. Why does a long microphone cable cause high–frequency losses?
7. How could dirt or dust cause a weak recording?
8. What are some of the factors that can cause poor high–frequency response?
9. What are some of the causes of poor low–frequency response?
10. What are some of the factors that can cause noise on a recording?
11. What are some of the problems that can be caused by defects in the mechanical system of a tape recorder?
12. What are some of the reasons for the tape to ride up and down on the capstan?
13. How is tape speed checked to best advantage?
14. How is the azimuth setting of a tape head accomplished?

6

BLACK-AND-WHITE TELEVISION TROUBLESHOOTING

6-1 GENERAL CONSIDERATIONS

Component defects in transistor television receivers result in picture and/or sound symptoms. Trouble symptoms are not necessarily unique, and a particular type of picture distortion can be caused by component defects in more than one section of a receiver. As an illustration, a no-picture and no-sound symptom could result from an "open" or "shorted" transistor in either the RF or the IF section of a receiver. It follows that localization procedures are required at the outset in this situation. Localization involves the same general approach that was explained for troubleshooting radio receivers. However, the approach is somewhat more complex in the case of a television receiver, due to the comparatively large number of sections, and because both a picture signal and a sound signal are processed by the receiver system.

A block diagram for a typical transistor television receiver is shown in Fig. 6-1. Note that the picture and sound signals proceed together through the signal channel to the video amplifier. At the video detector, the sound signal is separated from the picture signal, and is processed through a separate channel to the speaker. Signal tracing through the video IF section requires the use of a demodulator probe and oscillo-

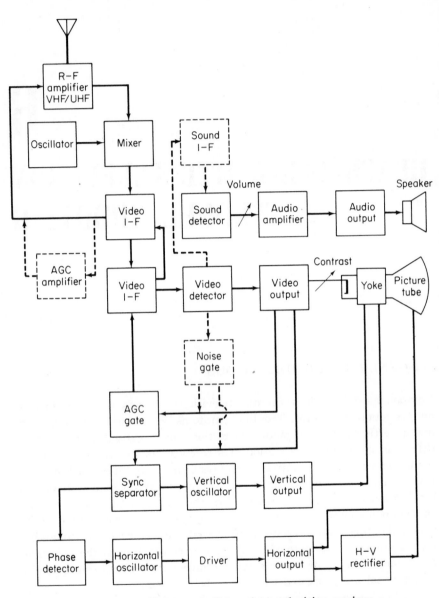

Figure 6-1 Block diagram for a small transistor television receiver.

scope, as explained in greater detail subsequently. Signal tracing through the video amplifier and the sound channel requires the use of a low-capacitance probe and wide-band oscilloscope that has full response up to 4.5 MHz.

Signal substitution tests can be made with an ordinary AM signal generator, provided that it has output up to 216 MHz. Although this frequency range covers only the VHF channels, harmonics can often be employed for signal substitution on the UHF channels. Stage gains are measured in the same basic manner as explained previously for radio receivers. Specified stage gains for a typical transistor–TV receiver are shown in Fig. 6–2. Note that there is normally a 6–dB difference in VHF gain from low–band to high–band operation. As seen in Fig. 6–3, the low VHF band encompasses frequencies from 54 to 88 MHz, whereas the high VHF band extends from 174 to 216 MHz.

Many technicians prefer to use specialized test–pattern generators in signal–substitution tests. A test–pattern display and the video signal waveform that produces it are shown in Fig. 6–4. Both test–pattern distortions and video waveform distortions provide important clues to various component defects. Details of test–pattern and waveform analysis are discussed in the subsequent topics. Pattern generators of the television–analyzer type also provide synchronizing and sweep waveforms for injection tests. An intercarrier sound signal is provided for signal–substitution tests in the sound channel.

6–2 PICTURE–CHANNEL TROUBLESHOOTING

Trouble symptoms resulting from picture–channel defects include no picture, no picture and no sound, weak picture, overloaded picture, smeared picture, negative picture, ringing, sync buzz, noisy sound, separated picture and sound, interference, picture pulling, and loss of sync. With reference to Fig. 6–1, it is evident that a component defect in the RF–amplifier section which stops the incoming signal will produce

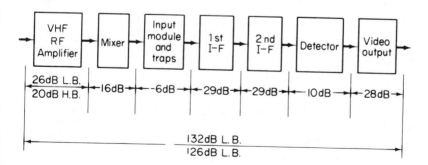

Figure 6–2 Normal stage–gain values for a small transistor–TV receiver.

P = PICTURE CARRIER FREQ. (MHz)
S = SOUND CARRIER FREQ. (MHz)

CHANNEL NUMBER	P	S	FREQ. LIMITS OF CHANNEL (MHz)
2	55.25	59.75	54–60
3	61.25	65.75	60–66
4	67.25	71.75	66–72
5	77.25	81.75	76–82
6	83.25	87.75	82–88
7	175.25	179.75	174–180
8	181.25	185.75	180–186
9	187.25	191.75	186–192
10	193.25	197.75	192–198
11	199.25	203.75	198–204
12	205.25	209.75	204–210
13	211.25	215.75	210–216
14	471.25	475.75	470–476
15	477.25	481.75	476–482
16	483.25	487.75	482–488
17	489.25	493.75	488–494
18	495.25	499.75	494–500
19	501.25	505.75	500–506
20	507.25	511.75	506–512
21	513.25	517.75	512–518
22	519.25	523.75	518–524
23	525.25	529.75	524–530
24	531.25	535.75	530–536
25	537.25	541.75	536–542
26	543.25	547.75	542–548
27	549.25	553.75	548–554
28	555.25	559.75	554–560
29	561.25	565.75	560–566
30	567.25	571.75	566–572
31	573.25	577.75	572–578
32	579.25	583.75	578–584
33	585.25	589.75	584–590
34	591.25	595.75	590–596
35	597.25	601.75	596–602
36	603.25	607.75	602–608
37	609.25	613.75	608–614
38	615.25	619.75	614–620
39	621.25	625.75	620–626
40	627.25	631.75	626–632
41	633.25	637.75	632–638
42	639.25	643.75	638–644
43	645.25	649.75	644–650
44	651.25	655.75	650–656
45	657.25	661.75	656–662
46	663.25	667.75	662–668
47	669.25	673.75	668–674
48	675.25	679.75	674–680
49	681.25	685.75	680–686
50	687.25	691.75	686–692
51	693.25	697.75	692–698
52	699.25	703.75	698–704
53	705.25	709.75	704–710
54	711.25	715.75	710–716
55	717.25	721.75	716–722
56	723.25	727.75	722–728
57	729.25	733.75	728–734
58	735.25	739.75	734–740
59	741.25	745.75	740–746
60	747.25	751.75	746–752
61	753.25	757.75	752–758
62	759.25	763.75	758–764
63	765.25	769.75	764–770
64	771.25	775.75	770–776
65	777.25	781.75	776–782
66	783.25	787.75	782–788
67	789.25	793.75	788–794
68	795.25	799.75	794–800
69	801.25	805.75	800–806
70	807.25	811.75	806–812
71	813.25	817.75	812–818
72	819.25	823.75	818–824
73	825.25	829.75	824–830
74	831.25	835.75	830–836
75	837.25	841.75	836–842
76	843.25	847.75	842–848
77	849.25	853.75	848–854
78	855.25	859.75	854–860
79	861.25	865.75	860–866
80	867.25	871.75	866–872
81	873.25	877.75	872–878
82	879.25	883.75	878–884
83	885.25	889.75	884–890

Figure 6–3 Frequency allocations for the VHF and UHF TV channels.

Figure 6–4 Standard test pattern and corresponding video waveform. (a) Display on screen of picture tube; (b) Video waveform displayed on screen of oscilloscope.

a no–picture and no–sound trouble symptom. In this situation, a high snow level will be displayed on the picture–tube screen when the contrast control is advanced, as illustrated in Fig. 6–5. The high snow level results from the facts that all of the picture–channel sections following the RF amplifier are operating normally, and the AGC section is also normal.

Note that if a no–picture and no–sound trouble symptom is caused by defective AGC action, no snow will be displayed on the picture–tube screen. This lack of snow results from the fact that both the front end and the IF strip are biased off in this situation. In other words, comparatively little snow is contributed by the latter portion of the picture channel, whereas any snow produced in the RF–amplifier or mixer stages is amplified by all of the succeeding stages, as is evident in Fig. 6–2. A

117

Figure 6–5 High snow level displayed on picture–tube screen.

snow display is produced by random noise voltages which are always present in operative circuits. When a medium–snow level is displayed, as shown in Fig. 6–6, it is likely that the signal is being stopped in the mixer stage. Confirming signal–tracing or signal–substitution tests are necessary, however, because some receivers have higher sensitivity than others, and produce a higher snow level when operating at maximum gain.

Since the IF section operates in the 40–MHz region, a demodulator probe must be used with a service–type scope in IF signal–tracing procedures. A typical demodulator probe is depicted in Fig. 6–7. The probe should be used with a comparatively high–gain scope, in order to obtain a usable display in low–level circuits such as the mixer and first IF

Figure 6–6 Medium snow level displayed on picture–tube screen.

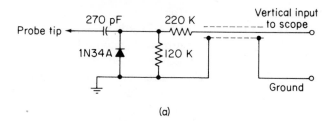

(a)

Frequency response characteristics

    RF carrier range: 500 kHz to 250 MHz
    Modulated-signal range: 30 to 5000 Hz
    Input capacitance (approx):   2.25 pF

Equivalent input resistance (approx.):

    At 500 kHz . . . . 25,000 Ω
         1 MHz . . . . 23,000 Ω
         5 MHz . . . . 21,000 Ω
       10 MHz . . . . 18,000 Ω
       50 MHz . . . . 10,000 Ω
     100 MHz . . . . .5000 Ω
     150 MHz . . . . .4500 Ω
     200 MHz . . . . .2500 Ω

Maximum input:

    ac voltage . . . . 20 rms V
                           28 peak V

(b)

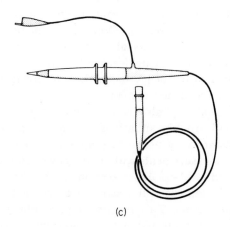

(c)

Figure 6–7 Low–capacitance probe for oscilloscope. (a) Typical probe circuitry; (b) Probe specifications; (c) Appearance of a low–capacitance probe.

stage. Note that a simple demodulator probe tends to detune and load the circuit under test. In turn, its chief utility is in showing the presence or absence of signal. Stage gain measurements should be made by signal injection. A test–pattern generator can be used as a signal source in either signal–tracing or signal–injection procedures. When a demodulator probe is used to trace a video signal, the oscilloscope should be operated at a 60–Hz or 30–Hz deflection rate. That is, a simple demodulator probe reproduces the vertical sync pulse without serious attenuation, whereas the horizontal sync pulse and camera signal are substantially attenuated.

A no–picture and no–sound symptom can be localized in the tuner section by means of either signal–tracing or signal–injection tests. As depicted in Fig. 6–8, a tuner employs a comparatively elaborate switching arrangement, and mechanical defects in the switches can cause a no–picture and no–sound symptom. Mechanical defects are particularly likely to occur in tuners that have been use for a long time. Sometimes contact trouble is caused by dust, dirt, or corrosion, and can easily be corrected with suitable cleaners and lubricants, as illustrated in Fig. 6–9. In other situations, worn or bent switch parts must be replaced. Technicians often send defective tuners to specialized repair centers for reconditioning because of the difficulties that are involved.

Apart from mechanical defects, the most common electrical defect in a tuner is a capacitor fault. As an illustration, if C15 (Fig. 6–8) "opens," the mixer stage will oscillate uncontrollably, and signal flow will be blocked at this point. That is, C15 operates as a neutralizing capacitor. Its action is most essential in low–band VHF operation, because the base and collector circuits resonate nearer the same frequency, particularly on the Channel 2 setting. Oscillation can be confirmed by means of DC–voltage measurements at transistor terminals, or with a frequency meter (wavemeter). A field–strength meter is generally used by service technicians as a frequency meter. Lab technicians utilize specialized high–accuracy frequency meters.

Defective transistors may also cause a no–picture and no–sound symptom in front–end trouble situations. As noted previously, DC–voltage measurements are very useful to pinpoint a defective transistor. Confirming resistance measurements with a low–power ohmmeter is helpful. Suspicion of a dead local–oscillator section can be checked as in Fig. 6–10. Note that balun damage will occur upon occasion in areas that have frequent thunderstorms. This hazard is particularly great if the television antenna is not protected with a lightning arrester. Even a minor lightning stroke that reaches a balun will burn out the coil windings. Ohmmeter tests will pinpoint a burned–out winding.

120

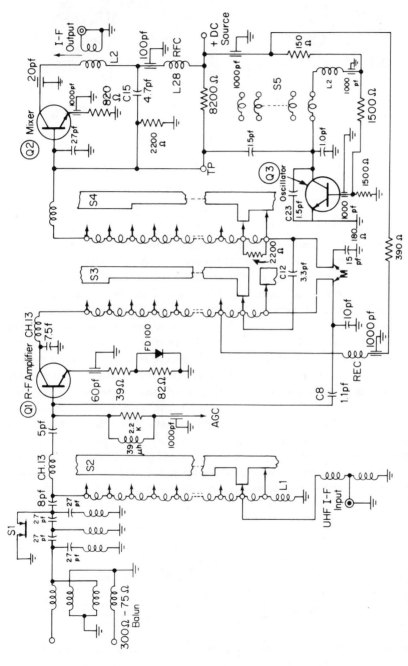

Figure 6–8 Configuration of a typical VHF tuner.

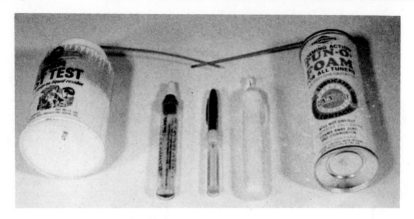

Figure 6–9 Switch contact cleaners and lubricants are utilized in tuner maintenance procedures.

A no–picture and no–sound trouble symptom can also be caused by a defect in the IF section. However, the AGC voltage should be measured, to eliminate the AGC section from suspicion. Sectionalization can be made to best advantage on the basis of signal–tracing or signal–substitution tests. A typical IF–amplifier configuration is depicted in Fig. 6–11. After it has been determined where the signal is being stopped, it is advisable to evaluate the trouble possibilities from the schematic diagram. As an illustration, if it were found that signal voltage is present at the input of the third IF transistor, but that no signal voltage is present at the collector, the prime suspects would be the base coupling capacitor, the neutralizing capacitor, or the transistor in the third stage. DC–voltage measurements will show whether either of the capacitors may be leaky or "shorted." A "bridging" test can be made to determine whether a capacitor is "open." If the transistor is defective, this condition will nearly always show up in DC–voltage measurements. Confirming resistance measurements can be made with a low–power ohmmeter.

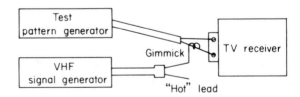

Figure 6–10 Output signal from a generator substitutes for a "dead" local oscillator.

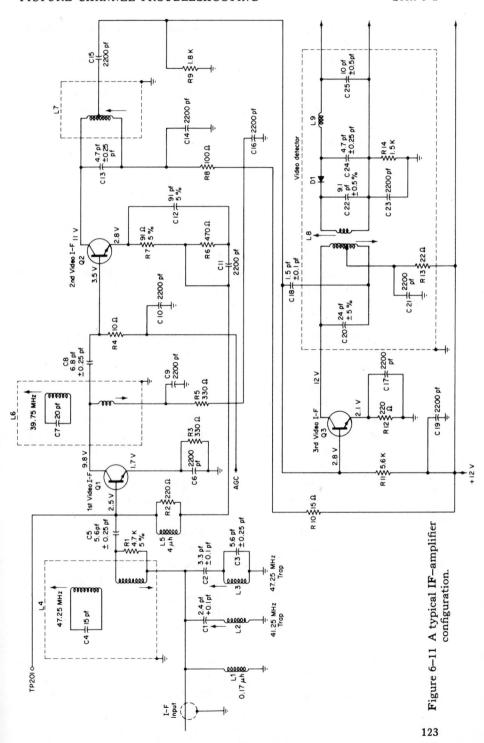

Figure 6-11 A typical **IF**-amplifier configuration.

It is evident that a no–picture and no–sound symptom can also be caused by a defect in the picture–detector stage, or possibly by a defect in the video–amplifier stage (depending upon the design of the particular receiver). With reference to Fig. 6–11, an "open" or "shorted" video–detector diode can cause signal stoppage. The configuration for a typical video-amplifier section is shown in Fig. 6–12. Note that a defect in the first video–amplifier stage will cause a no–picture and no–sound symptom, whereas a defect in the video–output stage will "kill" the picture only. Again an "open" coupling capacitor to the sound–takeoff transformer will "kill" the sound only. An oscilloscope and low–capacitance probe should be used to pinpoint the branch circuit that is causing the trouble symptom(s).

In the case of picture–tube failure, there will usually be no raster. However, an occasional defect occurs in a picture tube that "kills" the picture although the raster contines to glow. In such a case, an oscilloscope will show that the video signal is present up to the output of the video amplifier. Statistically, defective capacitors are the most common cause of video–section trouble symptoms. The video–output transistor is the next most likely suspect, because it operates at comparatively high voltage and high power. If a video–output transistor is replaced, it is essential to make certain that it is firmly secured to its heat sink (if used). Overheating is the most usual cause of power–transistor failure. Therefore, the base–cathode bias should be checked as a matter of routine when a video–output transistor is replaced.

A weak-picture symptom is generally caused by marginal failure instead of catastrophic failure, of the same components that are involved in a no–picture symptom. An overloaded (muddy and filled–up) picture can be caused by marginal leakage in a coupling capacitor that biases the associated transistor into its excessive–gain region. However, a component defect in the AGC section can cause the same picture symptom, and the AGC action should be checked at the outset. In case of doubt, the AGC line can be clamped at normal potential by means of battery bias or an override bias box. A smeared–picture symptom results from seriously distorted frequency response. As an illustration, in case the 10–μf decoupling capacitor in the video–detector output circuit (Fig. 6–12) loses capacitance or "opens," picture smearing results. The same symptom can be caused by regeneration in the IF amplifier, or by IF misalignment.

Negative–picture symptoms, often accompanied by 60–Hz sync buzz, are caused by defective transistors or incorrectly biased transistors in comparatively high–level stages, by regeneration due to misalignment or "open" neutralizing capacitors, or by incorrect component replace-

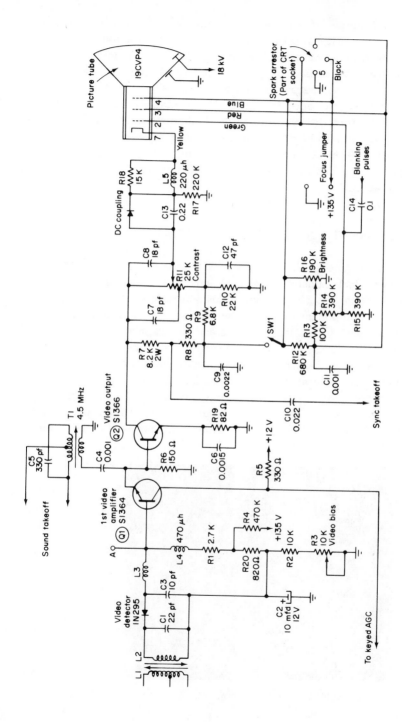

Figure 6–12 Configuration for a typical video–amplifier section.

ment. For example, if a replacement video–detector diode is connected with reverse polarity into the circuit, picture reproduction will be negative, as illustrated in Fig. 6–13. Picture pulling or sync loss, as shown in Fig. 6–14, can be caused by picture–channel defects such as incorrect biasing of the video–output transistor. In this situation, the sync pulses are compressed or clipped from the video signal. An oscilloscope check will show whether the sync pulses are being attenuated or "sliced off." This trouble can also be caused by IF or mixer regeneration, and may be accompanied by separated picture and sound. That is, when the fine-tuning control is set for best picture reproduction, the sound is weak, noisy, or absent.

6-3 SYNC–SECTION TROUBLESHOOTING

Symptoms of sync–section defects include loss of vertical and/or horizontal sync lock, unstable sync action, picture pulling, or vertical "bounce." Thus, the picture symptoms illustrated in Fig. 6–14 can be caused by a component defect in the sync section. Localization is made by checking waveforms with a scope and low-capacitance probe. As an illustration, the three waveforms and their specified peak–to–peak amplitudes shown in Fig. 6–15 provide key clues to circuit action. After a malfunctioning stage or branch has been localized, DC–voltage and resistance measurements generally suffice to pinpoint the defective component. The same approach is employed in troubleshooting the horizontal–AFC section, exemplified in Fig. 6–16. Note that horizontal–AFC diodes

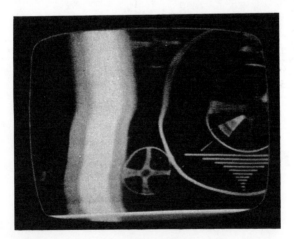

Figure 6–13 Example of a negative picture.

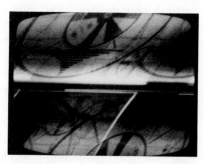

Figure 6–14 Picture trouble symptoms caused by attenuation of sync pulses. (a) Example of picture pulling; (b) Loss of horizontal sync; (c) Loss of vertical sync; (d) Loss of both horizontal and vertical sync.

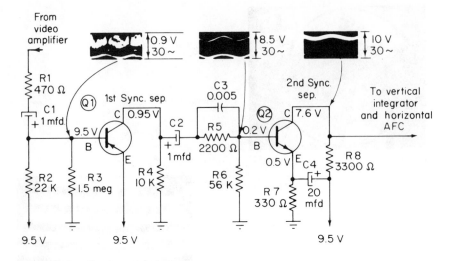

Figure 6–15 Configuration and normal waveforms for a typical sync–separator section.

should have a good front–to–back ratio, and should also be well matched. In general, diodes must be checked out–of–circuit to obtain meaningful test results. The reason for this requirement is exemplified in Fig. 6–17, which shows how shunt resistance masks front–to–back ratio test results.

6–4 SWEEP–SECTION TROUBLESHOOTING

Picture symptoms resulting from component defects in the vertical–sweep section include no vertical deflection, inadequate picture height, vertical nonlinearity, poor interlace, unstable vertical–sync lock, and keystoned raster. A typical vertical–sweep configuration is shown in Fig. 6–18. As in most vertical–oscillator designs, a blocking oscillator is employed. Note that diode M1 protects Q1 against inductive collector–voltage surges, which would break down the collector junction. Negative feedback is provided from the vertical output to the driver input by R20, to linearize the deflection waveform. Note that the vertical integrator is comprised of R1, R2, C1, and C2. If an integrating capacitor becomes defective, vertical sync–lock becomes impaired, with possible loss of interlace. An oscilloscope and low–capacitance probe are preferred for analysis of vertical–section circuit action.

Vertical nonlinearity and/or inadequate vertical height are commonly caused by defective electrolytic capacitors, such as C4, C5, C6,

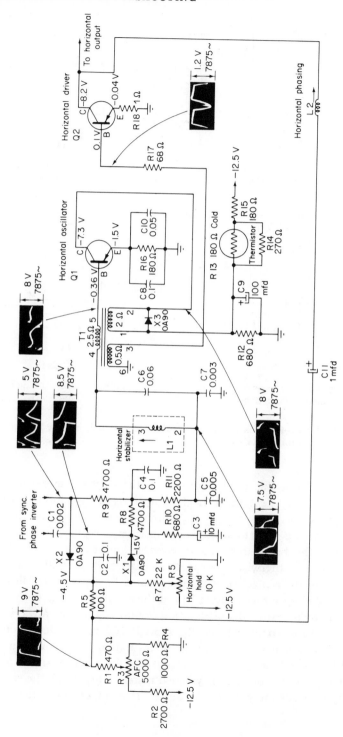

Figure 6–16 Configuration and normal waveforms for a horizontal–AFC section.

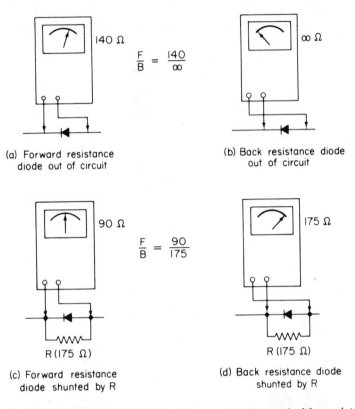

Figure 6–17 Front–to–back ratio test results are seriously masked by resistance in shunt to a diode.

C8, C9, or C10 in Fig. 6–18. However, if the electrolytic capacitors are not defective, the next most likely culprit is a deteriorated vertical–output transistor. DC–voltage measurements are most useful in this situation. In case a vertical–output transistor is replaced, it is essential to measure the base–emitter bias voltage, and to adjust this bias to normal value. Otherwise, the vertical–output stage will not operate properly, and the replacement transistor may be short lived. Bias stability is provided by a thermistor (R15 in the example of Fig. 6–18).

Trapezoidal raster distortion (keystoning) is nearly always caused by shorted turns, or a completely shorted coil in the deflection yoke. This trouble is generally the result of insulation breakdown and charring, although solder splashes or stray "whiskers" of stranded wire are occasionally responsible. If a keystoned raster is displayed, the technician generally replaces the yoke at the outset. It is important to use an exact

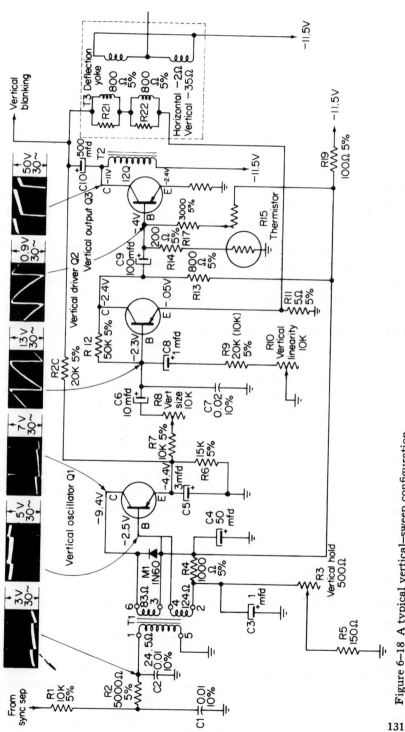

Figure 6–18 A typical vertical–sweep configuration.

replacement, inasmuch as a mismatch between the yoke and the vertical–output system, or a mismatch between the yoke and the picture tube will result in picture trouble symptoms, in spite of the fact that none of the components is defective.

Next, consider the horizontal–sweep configuration depicted in Fig. 6–19. This arrangement employs a gate–controlled switch (four–layer semiconductor device), which handles high peak power to better advantage than a bipolar power transistor. A gate–controlled switch is triggered into conduction by a gating pulse. It can then be triggered into nonconduction by applying an oppositely–polarized pulse to its base. In the case of a no–horizontal–sweep symptom, an oscilloscope and low–capacitance probe should be used to check the driving pulse. If this pulse has normal amplitude and fast rise, the trouble will be found in the output section, and not in the driver section. Other waveforms in the horizontal–output circuitry have somewhat limited usefulness in most trouble situations, due to the interacting circuit branches. Accordingly, component substitution tests are generally made. As in other receiver sections, capacitor defects are the most likely troublemakers.

Figure 6–19 exemplifies a hybrid arrangement, in which a high–voltage rectifier tube is utilized. Quite a few receivers of recent design employ a stack of semiconductor diode rectifiers instead of a tube, as seen in Fig. 6–20. This diode stack operates as a voltage multiplier, with a dozen steps of increasing voltage. When replacing a high–voltage rectifier stack, it is essential to observe the voltage rating. That is, a 6–kv stack, for example, should never be operated at 8 kv. The diodes are very likely to break down if their rated peak–inverse voltage is exceeded. Another practical consideration concerns the rated current capability of the semiconductor devices in a horizontal–output system. For example, it is hazardous to check for the presence of high voltage by arcing the high–voltage terminal to the chassis. Even momentary current overloads can damage devices that normally operate near their rated peak–power value.

6–5 ALIGNMENT PROCEDURES

Unless the tuned circuits in a TV receiver work together properly as a team, various difficulties will occur. As an illustration, when stagger–tuned IF stages resonate too near the center frequency, picture resolution (detail) is poor although the gain is high. Note that this operating condition is actually an advantage in fringe–area reception, due to the poor signal–to–noise ratio. In the event that IF stages are resonated at

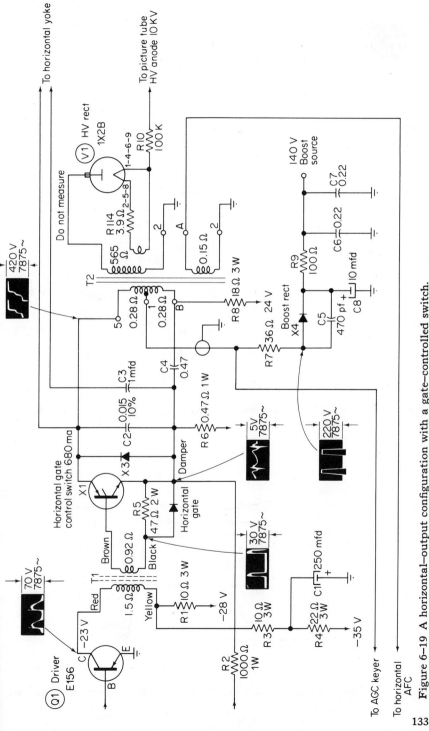

Figure 6–19 A horizontal–output configuration with a gate–controlled switch.

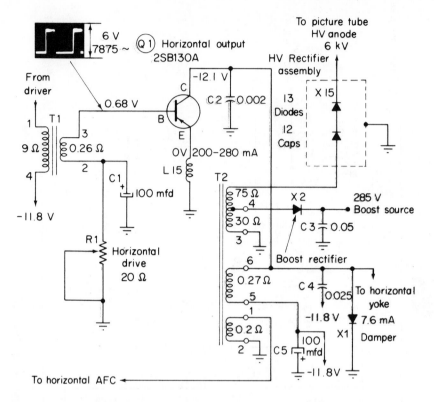

Figure 6–20 A stack of semiconductor diode rectifiers is employed in this HV rectifier assembly.

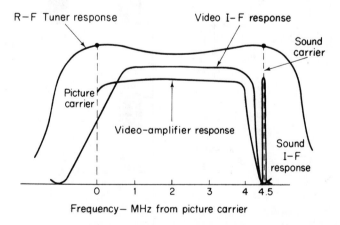

Figure 6–21 How the tuned circuits in a TV receiver operate together as a team.

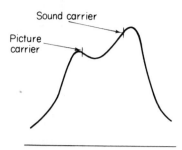

Figure 6–22 Example of a VHF frequency response curve with picture- and sound-carrier markers.

the same frequency, it is likely that the IF section will oscillate uncontrollably, with the trouble symptom of a "dead receiver." Figure 6–21 shows how the various signal–channel sections are related in frequency for normal operation.

Figure 6–22 illustrates the appearance of a frequency response curve for a VHF tuner, with markers at the picture– and sound–carrier points. A sweep generator, marker generator, and oscilloscope are utilized. The steps that are followed in a typical VHF tuner–alignment procedure are exemplified in Fig. 6–23. Note that compromise alignment adjustments are sometimes required in order to obtain acceptable frequency response on all VHF channels. The steps that are followed in a typical IF alignment procedure are exemplified in Fig. 6–24. The IF center frequency in this example is 25.25 MHz, as employed in various small solid–state TV receivers. Larger receivers generally use an IF center frequency of 45.25 MHz. Alignment procedures should be followed closely as specified in the applicable receiver service manual.

QUESTIONS

1. What are the requirements of the AM generator for it to be used for signal tracing in the video sound section?
2. If the RF amplifier is not operating, why does a high snow condition appear when the contrast is advanced to maximum?
3. Why is there no snow, no picture, and no sound when the AGC action is defective?
4. What causes a snow display?
5. What defects would cause a no–picture and no–sound symptom?

Oscillator adjustments

The individual oscillator slugs are accessible one at a time through a hole in the front of the tuner.
Set the fine tuning to the center of its range and adjust oscillator for best picture and sound on each active channel.

RF and Mixer Alignment

Connect the synchronized sweep voltage from the sweep generator to the horizontal input of the oscilloscope for horizontal deflection.
Use 10 MC sweep unless otherwise noted.
Connect a variable bias to the RF AGC line at point ⓣ. Adjust bias to obtain response curve which shows no indication of overloading.

	Sweep generator coupling	Sweep generator frequency	Marker generator frequency	Channel	Connect scope	Adjust	Remarks
1.	Across antenna terminals with 120 Ω in each lead.	213 MC	211.25 MC 215.75 MC	13	Vertical input to point ⓤ, low side to ground	A201, A202, A203, A204	Adjust for maximum gain and symmetry of response similar to fig. 201 with markers as shown.
2.	"	See chart	See chart	12 thru 2	Vertical input to point ⓤ, low side to ground		Check response on all channels and make compromise adjustments of A201, A202, A203 and A204 if required.

Sound   Video

Fig. 201

Channel and Frequency chart

Sweep generator frequency	Marker generator frequency	Channel	Sweep generator frequency	Marker generator frequency	Channel	Sweep generator frequency	Marker generator frequency	Channel
57 MC	55.25 MC 59.75 MC	2	85 MC	83.25 MC 87.75 MC	6	195 MC	193.25 MC 197.75 MC	10
63 MC	61.25 MC 65.75 MC	3	177 MC	175.25 MC 179.75 MC	7	201 MC	199.25 MC 203.75 MC	11
69 MC	67.25 MC 71.75 MC	4	183 MC	181.25 MC 185.75 MC	8	207 MC	205.25 MC 209.75 MC	12
79 MC	77.25 MC 81.75 MC	5	189 MC	187.25 MC 191.75 MC	9	213 MC	211.25 MC 215.75 MC	13

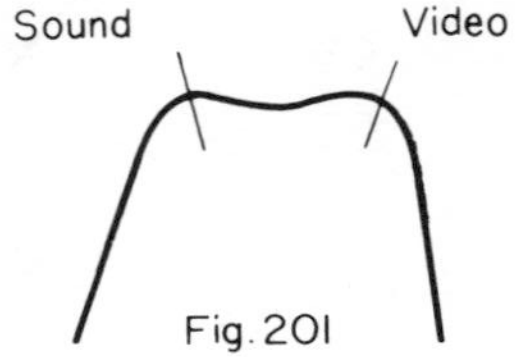

Tune to a UHF station and adjust UHF IF input coil for best picture and sound.

Figure 6-23 Typical VHF tuner-alignment procedure.

Use an isolation transformer and maintain voltage at 117 volts. Allow a 20-minute warm-up period for the receiver and test equipment
Suggested alignment tools : A1 thru A11 .... General Cement #9087.... Walsco # 2528
                            Mixer collector coil ...... General Cement #8284... Walsco # 2584

Video if alignment

Connect the synchronized sweep voltage from the sweep generator to the horizontal input of the oscilloscope for horizontal deflection.
Use only enough generator output to provide a usable indication. Note: response may vary slightly from those shown.
Connect a variable bias supply to the IF AGC line (point ⓐ) and adjust to obtain a response curve which shows no indication of overload
Disable oscillator section of mixer-osc. Set the channel selector to any non-interfering channel.

Indicator	Generator coupling	Sweep generator frequency	Marker generator frequency	Adjust	Remark
1 Connect DC probe of a VTVM thru a 47K resistor to point ⓐ Common to ground.	Connect high side thru a 100pf capacitor to point ⓑ on Tuner. Low side to ground		22.25 MC 28.25 MC	A1 A2	Adjust for minimum.
2 "	"		25.5 MC 26.3 MC 23.6 MC 24.3 MC	A3 A4 A5 Mixer Collector Coil	Adjust for maximum.
3 Connect vertical input of a scope to point ⓐ Low side to ground.	"	25 MC (10 MC sweep)	22.25 MC 23.50 MC 25.25 MC 26.75 MC 28.25 MC		Check for maximum gain and symmetry of response with markers as shown in figure 1. In order to obtain a proper response, it may be necessary to slightly retouch A3, A4, A5 and mixer collector coil.
4 Connect DC probe of a VTVM to a point ⓑ Common to ground.	"	44 MC (10 MC sweep)	26.75 MC	A6	Adjust for maximum.

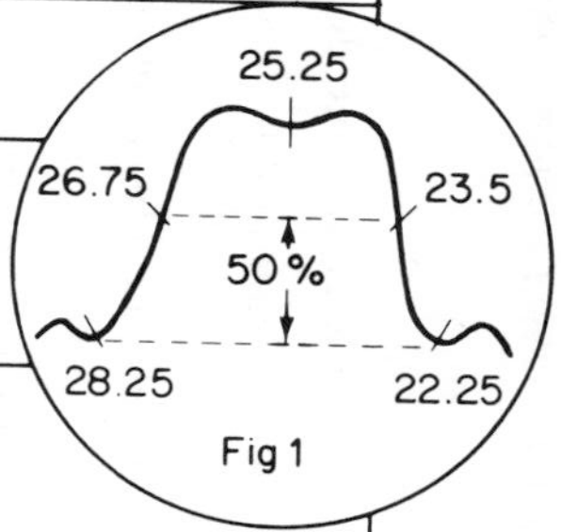

4.5 MC Trap alignment
Tune in a strong TV signal and set the contrast at maximum. Adjust the fine tuning until a beat pattern is visible on the screen.
Adjust A11 for minimum beat interference

Fig 1

Figure 6–24 Typical **IF** alignment procedure.

137

6. How are most defective tuners repaired?

7. What causes balun damage and how is damage determined?

8. In Fig. 6–11, what would be the symptoms if the video–detector diode were open?

9. What is the first thing that you should check when there is an overloaded picture symptom?

10. What are the usual causes of negative–picture symptoms?

11. What are the symptoms of sync-section defects?

12. What are the picture symptoms resulting from component defects in the vertical–sweep section?

13. What is the most common cause of keystoning of the picture?

14. What are the most likely circuit defects in the horizontal section?

7

TELEVISION CAMERA AND VIDEO TAPE RECORDER TROUBLESHOOTING

7–1 GENERAL CONSIDERATIONS

Many types of television cameras are in use by hobbyists, businessmen, industry, public service organizations, institutions, and entertainment centers. It is instructive to consider troubleshooting procedures for a typical closed–circuit television camera. Figure 7–1 illustrates the appearance of a closed–circuit TV camera. It utilizes a vidicon camera tube and associated deflection yoke, video amplifier, deflection circuits, sync waveshapers, modulated VHF oscillator, and power supply. The schematic diagram for the receiver is shown in Fig. 7–2. Details of circuit action are discussed subsequently.

A 1-inch vidicon tube drives the video amplifier which has a bandwidth of 10 MHz. The video–frequency output has an amplitude of 1 volt peak–to–peak into a 75–ohm load. Random sync interlace is employed. The modulated VHF output is tunable over channels 2 through 5, and has an amplitude of 50,000 microvolts into a 300–ohm load. Center resolution is 500 lines, and corner resolution is 400 lines. A three-position lens turret is provided, and the focusing mount is adjustable from 2 feet to infinity. Except for the focus and iris adjustments, and the off–on power switch, no external controls are provided. The camera can

Figure 7-1 A high–quality closed–circuit television camera (*Courtesy of* Diamond Electronics Co.).

be used to energize either a standard black–and–white **TV** receiver, or a video monitor. More detailed reproduction is obtained with a monitor, due to the limited bandwidth of a conventional **TV** receiver.

7–2 OPERATING NOTES

Closed–circuit **TV** cameras utilize transformer–operated power supplies, so that there is no abnormal shock hazard either to the operator or to the troubleshooter. On the other hand, if the output cable from the camera is connected to a **TV** receiver or to a video monitor that has a "hot chassis," the camera frame will then be connected to one side of the power line, and a definite shock hazard will be present. The video output cable from a CCTV camera should be terminated by a 75–ohm load. Note that no more than 500 feet of output cable should be employed, or the signal will become objectionably attenuated. If modulated

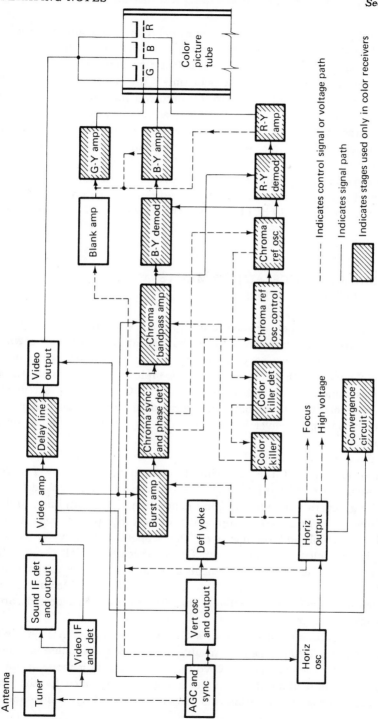

Figure 7–2 Schematic diagram for the receiver.

VHF output is being used, the 75-ohm cable may be connected directly to the 300-ohm input terminals of a television receiver without a matching pad, and the image degradation will not be pronounced.

Image reproduction will be poor unless the camera lens is carefully focused, and the iris adjusted for optimum contrast under the prevailing lighting conditions. If too much light is admitted into the vidicon tube, the contrast will be abnormally high and picture detail will be degraded. On the other hand, too little light admittance results in subnormal contrast and noise interference in the image. Note that a vidicon tube will be damaged immediately if it is directed at the sun or any intense source of light without a suitable filter and reduced iris opening. A cap is provided, which should be placed over the lens when the camera is not in use.

7-3 MAINTENANCE ADJUSTMENTS AND TROUBLESHOOTING

When a CCTV camera is used to energize a TV receiver with modulated VHF output, local interference (co-channel interference) may occur. In such a case, it is necessary to adjust the camera for operation on another channel. With reference to the example under discussion, the top cover is removed from the camera, and the Channel-Adj. screw (Fig. 7-3) is turned to provide optimum reproduction on the particular channel to which the receiver is set. The top cover is then replaced on the camera.

From the viewpoint of the troubleshooter, a CCTV camera is considered in terms of the following functional sections: power supply, focus-coil current regulator, vertical deflection, horizontal deflection, blanking generation, synchronizing generation, video amplifier, and RF modulator. A minimum complement of test equipment consists of a VOM or TVM and oscilloscope. The scope should have a vertical-amplifier frequency response from DC to at least 10 MHz. In addition, the standard bench test equipment used in TV service shops will be helpful. Fixed capacitors are the most common troublemakers; if a capacitor checker is not available, substitution tests of suspected capacitors may be made.

In each functional section of a CCTV camera, there are certain key check points. To localize any trouble symptom, these points are tested first. In the power supply, a DC voltage measurement is made at point G (Fig. 7-2). A reading of +18 volts is normally obtained for the example under consideration. Abnormal or subnormal voltage values indicate

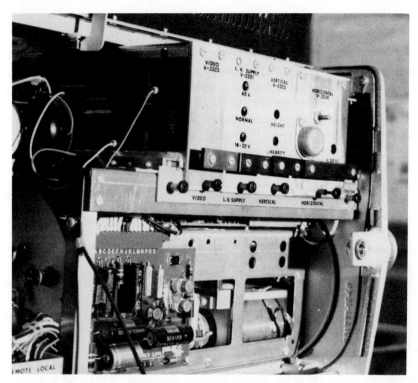

Figure 7-3 Inside view of the CCTV camera, showing location of the Channel–Adjustment control (*Courtesy of* Ampex Corp.).

trouble in the power supply. In the horizontal–deflection section, an oscilloscope is used to check the waveform at the collector of Q11 (Fig. 7-2). Negative pulses with a width of 10 microseconds, spaced 63.5 microseconds apart, and with an amplitude of 65 volts peak–to–peak normally are observed. This test requires a scope with triggered sweeps and a calibrated time base. Abnormal waveshapes or incorrect peak–to–peak voltage values indicate trouble in the horizontal–deflection section.

In the vertical–deflection section, an oscilloscope is applied to check the waveform at the collector of Q13 (Fig. 7-2). Normally, a negative-going 60–Hz sawtooth with an amplitude of 4 volts peak–to–peak is displayed in the example under consideration. In the sync section, an oscilloscope is applied to check the waveform at the collector of Q9 (Fig. 7-2). A combination horizontal–and–vertical sync waveform is normally observed. The vertical pulse should have a width of approximately 300 microseconds, the horizontal pulse 5 microseconds. These pulses are normally positive–going and have an amplitude of about 18 volts peak–to–peak. An abnormal waveshape or incorrect amplitude indicates trouble in the sync section.

The blanking waveform is checked at the collector of Q8 (Fig. 7–2). A combination horizontal–and–vertical blanking waveform is normally displayed. The vertical pulse should be approximately 1300 microseconds wide, the horizontal pulse 10 microseconds wide. They are normally positive–going with an amplitude of about 18 volts peak–to–peak. An abnormal waveform or incorrect amplitude indicates trouble in the sync section. In the video–amplifier section, an oscilloscope is applied at the emitter of Q7 (Fig. 7–2). A composite video signal with negative–going sync should be observed, with an amplitude of approximately 1 volt peak–to–peak. The sync pulses normally have an amplitude of about 0.4 volt peak–to–peak. Distorted waveforms or incorrect amplitudes point to trouble in the video amplifier.

Input and output checks are made in the modulator section. We check the input to the modulator by observing the video signal at J1 (Fig. 7–2). Then, the scope probe is moved to J2, and the output waveform is checked. This output waveform should have an amplitude of about 50 millivolts, with negative–going sync. A distorted waveform and/or incorrect peak–to–peak voltage indicates trouble in the modulator section. After a trouble symptom has been localized to a particular section of the camera system, the technician then proceeds to make tests that will pinpoint the defective component. These tests generally involve DC voltage and resistance measurement.

7–4 IDENTIFICATION OF DEFECTIVE COMPONENTS

Power Supply. In the case of a defective power supply, it is good practice to first measure the AC voltage at the power–line outlet. If the power–line voltage is normal, we next check for a blown fuse (F1 in Fig. 7–2). If the fuse is not blown, it is advisable to remove the cover from the camera, plug in the line cord, turn on the power switch, and observe the vidicon tube to determine if its filament is glowing. If the tube is dark, its filament–supply voltage is measured; this is normally 6.3 volts AC. In case the supply voltage is normal, the tube filament is probably burned out. On the other hand, weak or zero supply voltage throws suspicion on the power transformer.

Power Transformer Circuit. With the camera line cord unplugged from the AC outlet, we first measure the resistance between the prongs on the plug. With the power switch turned on, a reading of 65 ohms is normally obtained in the example under discussion. A reading of infinity can be caused by a defective line cord, defective switch, defective fuse holder, or burned–out primary winding. A reading substantially less than 65 ohms points to a partial short–circuit, such as short–circuited layers in the primary winding. If the primary circuit checks out satisfactorily, the technician proceeds to check the output voltages on the secondary side of the transformer (Fig. 7–2). If a secondary voltage is zero, the associated winding is probably open; if low, there may be shorted turns or layers in the winding.

If all the secondary output voltages are correct, a DC voltage measurement is made next at the positive terminal of C40 (Fig. 7–2). A zero voltage reading indicates that CR13 is probably defective. On the other hand, if the reading is normal (+26 volts), proceed next to point G and measure the voltage to ground (+18 volts). If the reading is zero, Q14 is probably defective; if the reading is the same as at the positive terminal of C40, Q14 is defective (short–circuited). Note that if the voltage at point G is abnormal and cannot be set at +18 volts by adjustment of R81, point–by–point measurements should be made through the regulator circuit. Specified DC voltages for this example are shown in Fig. 7–4. If it appears that a transistor is defective, a substitution test is recommended.

In the event that the 18–volt regulated power supply checks out

Transistor dc voltages

Transistor	Emitter	Base	Collector
Q1	+3.7 Vdc	+3.35 Vdc	+15 Vdc
Q2	0 (zero)	+0.55	+4.6
Q3	+4.0	+4.6	+8.8
Q4	0 (zero)	+0.65	+12.5
Q5	+0.75	+1.15	+11.0
Q6	+5.5	+6.2	+12.0
Q7	+11.5	+12.0	+18.0
Q8	0 (zero)	+0.3	+3.0
Q9	0 (zero)	+0.65	+1.2
Q10	Refer to text		
Q11	Refer to text		
Q12	0 (zero)	−1.8	+8.4
Q13	+0.4	+1.1	+7.3
Q14	+18	+18.5	+26
Q15	+6	+6.6	+18.5
Q16	+5.4	+6	+13

Figure 7–4 Specified transistor DC voltages for the configuration of Figure 7–2.

145

satisfactorily, the technician turns his attention to the 300–volt supply. The voltage at the positive terminal of C37 (Fig. 7–2) is measured. The reading is normally 340 volts. A zero reading indicates that CR9 and CR10 are probably defective; however, a short–circuit condition might be present. Next the voltage at point B is measured. Then the voltage at point A is measured. A zero reading indicates that R72 is probably defective. If all of the power–supply voltage checks are satisfactory, the power supply is cleared from suspicion, and the next logical trouble source is analyzed.

Horizontal Deflection. An overall check of the horizontal–deflection circuit (Fig. 7–2) is made by observing the waveform at the collector of Q11. We normally find a negative–going pulse with a 15,750–Hz repetition rate, a width of 10 microseconds, and an amplitude of approximately 65 volts peak–to–peak. To correct any error in repetition rate, R56 is adjusted. If adjustment cannot be made, Q10 is probably defective. An abnormal or subnormal pulse amplitude is corrected by adjusting R58. Inability to obtain the correct amplitude indicates that Q11 is defective. Transistor defects almost always show up as incorrect terminal voltages. In this example, the normal operating voltages for Q10 and Q11 are as follows:

Q10	Emitter, +7v	Base 1, +0.6v	Base 2, +16v
Q11	Emitter, +8.8v	Base, +8.8v	Collector, +4.2v

An oscilloscope provides a confirming or supplementary test. A sawtooth waveform with an amplitude of about 7 volts peak–to–peak should appear at the emitter of Q10 (Fig. 7–2). If no waveform is present, Q10 is not oscillating. A positive pulse with a width of 10 microseconds and an amplitude of approximately 3 volts peak–to–peak also should appear at base 2 of Q10. If this waveform is absent, Q10 is probably defective. It is also important to make resistance measurements to verify continuity of the deflection yoke and of coils L5 and L6.

Vertical Deflection. An overall check of the vertical–deflection circuit (Fig. 7–2) is made by observing the waveform at the collector of Q13. A negative sawtooth with a 60–Hz repetition rate and an amplitude of about 4 volts peak–to–peak is normally displayed. If the waveform is distorted and/or has incorrect amplitude, it is probable that there is trouble in the vertical–deflection section. The next logical step is to check the drive at Q13. A positive sawtooth with an amplitude of approximately 0.5 volt peak–to–peak is normal. The waveform at the collector of Q12 should also be checked. This is a negative–going pulse with a width of about 1300 microseconds and an approximate amplitude of 7 volts peak–to–peak.

In the case of no output from Q12 (Fig. 7–2), the base drive waveform should be observed. It should be a typical base–blocking ("cash-register") waveform at about 4.5 volts peak–to–peak. If this signal is absent, the technician proceeds to check for AC input at D and E, and to the primary of blocking transformer T2. To pinpoint a defective component, the DC voltages noted in Fig. 7–4 are employed as a guide. Note that the vertical sweep amplitude is adjusted by setting R49 for a test–pattern aspect ratio of 4:3. Vertical centering is adjusted by setting R55 to the point that the DC voltages from either side of the vertical-deflection coil to ground are equal.

Synchronizing Circuitry. Synchronization trouble is analyzed by first observing the combined sync waveform at the collector of Q9 (Fig. 7–2). If the vertical pulse is missing, check for vertical–frequency input at the junction of R45 and C21. On the other hand, if the horizontal pulse is missing, check for horizontal–frequency input at the junction of R42 and R43. Check also for the combined sync waveform at the base of Q9. The waveform amplitude should be about 1.8 volts peak–to–peak. To pinpoint a defective component, the DC voltages noted in Fig. 7–4 are used as a guide.

Blanking Circuitry. Trouble symptoms involving the blanking function are first analyzed by observing the combined blanking waveform at the collector of Q8 (Fig. 7–2). If the vertical pulse is missing, proceed to check for the vertical–input waveform at the junction of R45 and C21. On the other hand, if the horizontal pulse is missing, check for the horizontal–input waveform at the junction of R42 and R43. Also check the combined waveform at the base of Q8. Its amplitude should be approximately 1.8 volts peak–to–peak. To pinpoint a defective component, refer to the DC voltage values specified in Fig. 7–4.

Video Amplifier. Troubleshooting the video amplifier is comparatively straightforward, because sync and blanking signals are injected at different points in the amplifier configuration. For example, if the sync signal only is present at J1, Q6 and Q7 are operating properly, since the sync signal is injected at the base side of Q6 (Fig. 7–2). Again, if blanking and sync pulses are present at the output of J1, stages Q4, Q5, Q6, and Q7 are operating properly. If blanking is present at the output of Q5, but is not present at the output of J1, diode CR3 is probably defective. If the technician places his finger near the case of Q2, considerable noise normally appears in the output. Note that the signal level at the output of the vidicon, or at the base of Q2, is comparatively low and cannot be measured without using a high–gain scope and isolating probe. It is easier to check at the collector of Q2, where the normal amplitude is about 20 millivolts. To pinpoint defective components, DC voltage measurements should be made and compared with the values specified in Fig. 7–4.

RF Modulator. It is not difficult to troubleshoot the RF modulator. If no output is present, check for video input to the modulator. Diodes CR1 and CR2 are likely to be defective. To confirm that the oscillator is operating, a DC voltage measurement at the junction of CR1 and CR2 should read approximately −0.38 volt. To pinpoint a defective component, make DC voltage measurements and compare with the values specified in Fig. 7–4.

7–5 GENERAL CONSIDERATIONS OF VIDEO TAPE RECORDER TROUBLESHOOTING

A video tape recorder can store a television program for playback at any time. Both sound and picture are included on the magnetic tape recording, and are ready for instant replay, if desired, because no time is required for processing. The more elaborate types of video tape recorders can store a color–television program. Slides may be recorded for still commercials, with sound provided by an announcer. Video tape recorders are employed for many TV programs and commercials because the material can be recorded at any convenient time, and then played back at any desired time. Figure 7–5 shows the appearance of a typical video tape recorder, with a monitor. A VTR may be utilized either for closed–circuit recording or for storing commercial broadcast programs.

Magnetic tapes used in video tape recorders may be from one–half inch to two inches wide. A tape moves at a speed from 3.75 to 9.6 inches per second. However, the recording and playback heads also move across the tape, so that the relative tape–to–head speed is comparatively high. For example, one type of VTR operates with a relative tape–to–head speed of 1500 inches per second, and can record frequencies up to 5 MHz. A widely used design employs helical recording, as depicted in Fig. 7–6. Two record/playback heads are mounted on a rotating drum and trace a diagonal curve across the tape. Frequencies up to 3.2 MHz are recorded by this design.

Figure 7–5 A typical video tape recorder and monitor, with a closed–circuit TV camera (*Courtesy of* Ampex Corp.).

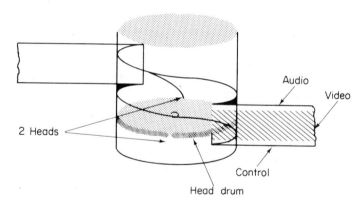

Figure 7–6 Helical method of video tape recording (*Courtesy of* Ampex Corp.).

7–6 TAPE TRANSPORT

As in a conventional tape recorder, the tape transport mechanism controls tape movement in forward, fast forward, and rewind operations. Figure 7–7 shows the basic features of transport. As the tape moves from the supply reel to the takeup reel, it passes the tension arm pin. This device senses the tape tension and operates to control braking pressure to

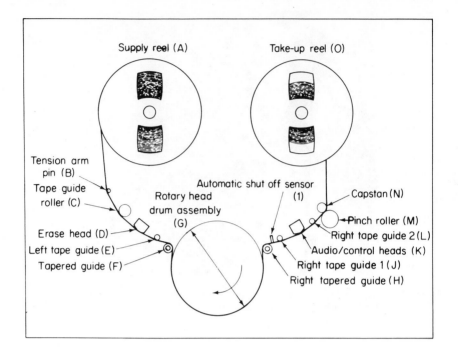

Figure 7–7 Basic features of video tape transport mechanism.

the supply reel as required to keep a uniform back tension on the tape. The tape then proceeds past the tape guide roller, the erase head, and the left tape guide. These guides ensure that the tape crosses the erase head at correct height. In addition, the left tape guide changes the tape angle slightly, to control its motion across the left tapered guide. This left tapered guide angles the tape downward as it begins its path around the rotary head drum assembly.

Since the video heads inside the assembly rotate in a horizontal plane in Fig. 7–7, they follow paths that slant across the width of the tape. Next, as the tape leaves the rotary head drum assembly at the right side, it is bent back into a horizontal direction by the right tapered guide. The tape then passes the automatic shutoff sensor and the right tape guide. Note that if the tape runs out, the automatic shutoff sensor moves toward the front of the machine and actuates a microswitch that cuts off the AC power. After passing right tape guide 1, the tape then presses against the audio/control heads and passes right tape guide 2. The audio track is recorded on the upper edge of the tape and control pulses are recorded on the lower edge. These control–track pulses provide a timing reference for the transport playback system. The tape is squeezed between the capstan and pinch roller, and fed finally to the takeup reel.

7–7 MECHANICAL MAINTENANCE

Noise in the picture during playback is usually caused by an accumulation of debris in one or both video heads. Sometimes half the picture may be noisy. In severe situations there is no video output. **Figure 7–8** shows the appearance of a video head assembly. To clean the heads, the tape is removed, and one of the heads is moved to the cleaning position located near the left tapered guide. The head can then be cleaned as depicted in **Fig. 7–9.** A cleaning tip is saturated with methanol, or

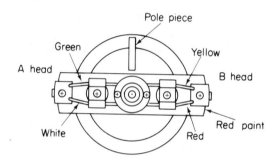

Figure 7–8 Appearance of a video head assembly.

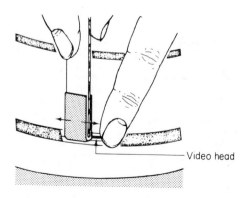

Figure 7–9 Video head is cleaned by rubbing horizontally.

equivalent, and the cleaning tip is rubbed across the head from side to side. The tip should not be rubbed vertically; damage could result. On the other hand, erase and audio/control heads are rubbed vertically while cleaning.

Intermittent dark horizontal lines in the picture are caused by noisy slip rings in the rotary–head drum assembly. The drum cover is removed and the slip rings are wet with a few drops of cleaning solution. Then the motor is turned on for 10 to 20 seconds, after which any excess fluid is wiped away. If slip–ring noise still persists, the rings may be cleaned directly by rubbing with a cleaning tip saturated with cleaning fluid. Bearing lubrication is required at periodic intervals to ensure against excessive wear. However, the sliding parts normally do not require lubrication. Oil should not be spilled carelessly around the transport mechanism, and any excess oil should be wiped off.

After approximately 500 hours use, the video heads may be sufficiently worn that picture reproduction is noticeably degraded. The heads must be replaced at this time. First the head–drum cover is removed. Then the brush is removed from the spring (see Fig. 7–10).

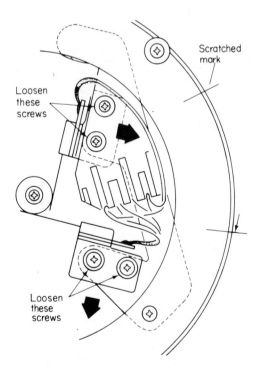

Figure 7–10 Procedure for video head replacement.

Scratch marks are made to ensure that the upper drum will be replaced correctly. The mounting screws are loosened, and the drum top is lifted back. This exposes the video head assembly (Fig. 7–8), and its mounting screws can be loosened. In turn, the assembly can be removed, and a replacement video head assembly installed. Brush pressure adjusting screws are turned to obtain approximately a 2–mm bend, as depicted in Fig. 7–11. It is important to center the brushes in the slip rings.

Video head dihedral adjustment is occasionally required (the two video heads must be exactly 180° apart). The method of dihedral adjustment is seen in Fig. 7–12. A special video alignment tape is used as a guide in resetting the adjustment screws. This test tape provides a special test pattern as shown in Fig. 7–13, which is observed on the screen of a monitor. In case dihedral adjustment is needed, it will be observed that there is more or less jitter at the top of the pattern, as depicted in A. When the position of the head (Fig. 7–12) is correct, the top of the picture is displayed normally, without any evidence of jitter.

If the tape tends to slip, the pinch roller pressure is probably incorrect. With reference to Fig. 7–14, there should be 0.1 mm clearance between the lower end of the pinch roller and the capstan shaft when the upper end of the pinch roller contacts the capstan. If necessary the pinch lever is bent at **A** with a pair of pliers to obtain the correct spacing. It is also essential to have the correct pressure between the pinch roller and the capstan. This is measured with a spring scale, as depicted in Fig. 7–14. When the roller is pulled back to just clear the capstan, the scale should read between 1.8 and 2.5 kg. Otherwise, the roller pressure spring should be replaced.

In many aspects, mechanical maintenance of video tape recorders

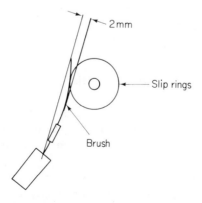

Figure 7–11 Adjustment of brush pressure.

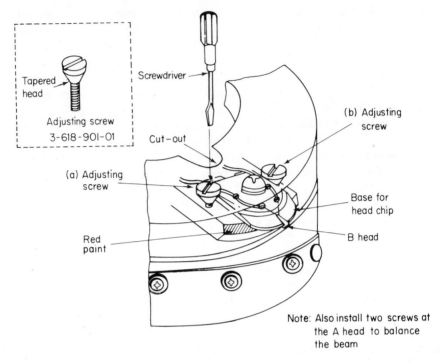

Figure 7–12 Video head dihedral adjustment.

can be compared with similar procedures for audio tape recorders. However, video recorders are somewhat more complex, as exemplified by the techniques of head adjustment and replacement. For this reason, it is helpful for the beginner to acquire a reasonable amount of experience in troubleshooting audio tape recorders before attempting to cope with video tape recorders. Expertise in this area requires a balanced combination of theoretical knowledge and practical experience.

7–8 ELECTRONIC SYSTEM TROUBLESHOOTING

Troubleshooting of the electronic system in a video tape recorder involves the same general principles that are used in servicing television receivers. Test equipment is also essentially the same. In the case of color video tape recorders, heterodyne functions may be employed in addition to video amplifiers. That is, the color subcarrier may be heterodyned down to a lower frequency, which is more easily recorded on tape. In turn, the color subcarrier is heterodyned back to its standard NTSC

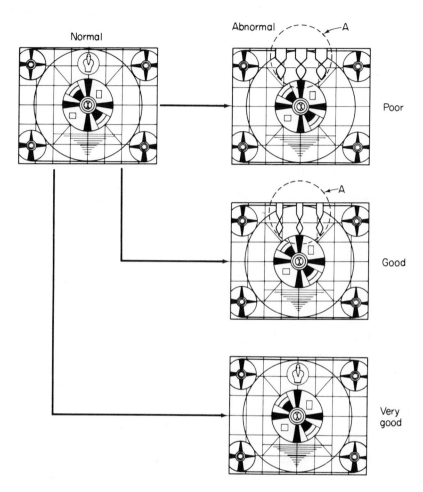

Figure 7–13 Test–pattern indication of normal and abnormal dihedral adjustments.

value on playback. Therefore, the video section of a color video tape recorder is more complex than the color section in a color–TV receiver.

Service data for video tape recorders are much the same as for television receivers. Waveforms and peak–to–peak voltages are specified throughout the system, and DC voltages are specified at transistor terminals and other key check points. Picture trouble symptoms will sometimes point to a particular circuit section which is defective, but in most cases an oscilloscope must be used to check the waveforms in various sections. In nearly all situations, a circuit or component defect will show up as a distorted and/or incorrect amplitude waveform. To pinpoint a defective component in a circuit, DC voltage measurements are made

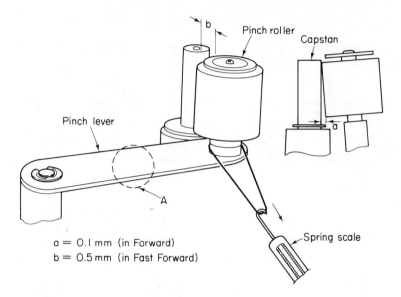

Figure 7–14 Pinch roller angle and pressure adjustment.

and compared with the values specified in the recorder service data. There is a marked trend to module design; this feature simplifies troubleshooting procedures considerably, because a suspected defective module can be quickly checked by means of a substitution test.

QUESTIONS

1. What are the effects of a variation of light on the quality of the picture from a vidicon camera?
2. What are the effects on a vidicon tube that is directed at the sun?
3. How can you remove co-channel interference from a CCTV system?
4. From the viewpoint of the troubleshooter, what are the functional blocks of a CCTV system?
5. What test equipment should a technician have to troubleshoot a CCTV camera?
6. In Fig. 7–2, where is the overall vertical–deflection circuit checked?
7. How is a defective component pinpointed in the circuit in Fig. 7–4?
8. What is the tape speed of a typical video recorder?
9. What causes intermittent dark horizontal lines in the picture?
10. Why are scratch marks made on the upper drum when replacing the video head?

8

COLOR–TELEVISION TROUBLESHOOTING

8–1 GENERAL CONSIDERATIONS

Fundamentally, a color–television receiver employs all of the conventional black–and–white circuitry, plus a chroma section with a color picture tube, as depicted in Fig. 8–1. In turn, all of the picture and/or sound trouble symptoms that occur in black–and–white receivers are confronted in color reception, plus trouble symptoms that are unique to

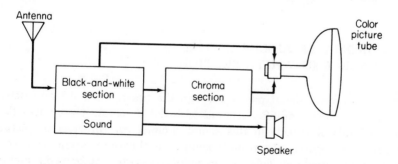

Figure 8–1 The fundamental plan of a color–television receiver.

chroma circuitry and color picture tubes. Color–TV receivers provide compatible operation (the receiver system automatically switches circuit sections as required to process color signals or black–and–white signals). As an illustration, during black–and–white reception, the chroma section depicted in Fig. 8–1 is disabled by electronic–switch action. This electronic switch is called a color killer, and is actuated by a basic color-signal component called the color burst.

Observe the block diagram for a solid–state color–TV receiver shown in Fig. 8–2. As explained in more detail subsequently, the incoming color signal has a conventional black–and–white component, plus a chroma component. The black–and–white component is called the Y signal. It proceeds through the tuner to the video–IF and detector section, and to the sound–IF and detector sections. From the picture detector, the Y signal proceeds through the video amplifier and delay line to the video-output section. Instead of being applied directly to the picture tube, the Y signal then proceeds through the color–demodulator and color–video sections, terminating at the cathodes of the color picture tube. Note that the Y signal is merely amplified through the color–demodulator and color–video sections. That is, the Y signal is not processed in these sections.

Meanwhile, the color burst has actuated the color killer, which has turned the chroma–bandpass amplifier on. The chroma component of the incoming color signal is amplified through the bandpass amplifier, and applied to the color demodulators. In addition, the chroma–reference oscillator has been synchronized by the color burst, and a reconstituted color–subcarrier voltage is applied to the color demodulators. This chroma–processing action decodes the incoming chroma signal, and forms the complete color signal. In turn, the cathodes of the color picture tube are driven by the combined Y and decoded chroma signals, and a color picture is displayed on the picture–tube screen. These topics are explained in greater detail subsequently.

8–2 CHROMA TROUBLES IN THE BLACK–AND–WHITE SECTION

Since the chroma signal passes through the tuner, video–IF and detector, and the video–amplifier sections with the Y signal, it is evident that various kinds of defects in this signal channel can affect color–picture reproduction. Among the trouble symptoms that may occur are: weak color and/or unstable color sync action, distorted color reproduction with possible unstable color sync action, poor color "fit," or complete

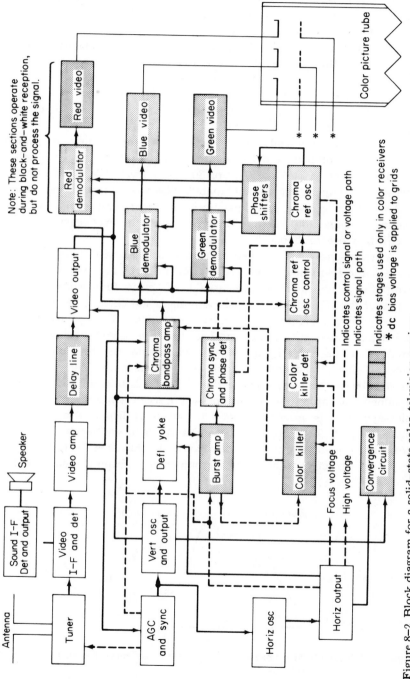

Figure 8–2 Block diagram for a solid–state color–television receiver.

loss of color with acceptable black–and–white reception. When the chroma signal is attenuated in passage through the signal channel, the reproduced color in the image is washed–out or pale, although the color–intensity control has been advanced to maximum. Component defects that attenuate the chroma signal also attenuate the color burst, with the result that color–sync action becomes impaired.

One of the common causes of a weak–color symptom is misalignment of the IF amplifier or the RF tuner. Figure 8–3 shows a normal frequency–response curve for an RF tuner. Note that the color subcarrier appears at practically the same level as the picture and sound carriers. If a component defect causes such frequency distortion that the color subcarrier is reduced 20 or 30 dB in level, it is evident that a weak–color symptom can be anticipated. The most common component fault is a leaky or open fixed capacitor in the tuner circuitry. Bypass and decoupling capacitors often seriously distort the frequency–response curve if they become "open."

A sweep–and–marker generator should be used with an oscilloscope to check the frequency response of an RF tuner or IF strip. Figure 8–4 exemplifies a typical IF frequency–response curve for a color receiver. Note that the IF section, bandpass amplifier, and chroma demodulators normally work together as a team. In turn, if an IF–amplifier component defect causes such frequency distortion that the chroma subcarrier is reduced 10 dB in level, the reproduction of color will be greatly weakened or "killed," and unstable color sync action is also likely to occur. The troubleshooting approach in this situation is basically the same as explained in Chapter 6 for black–and–white receivers. Observe that when color–sync lock is broken, a color–bar pattern is displayed as shown in Fig. 8–5(a). If both color–sync lock and black–and–white–sync lock are broken, a color–bar pattern is displayed as seen in Fig. 8–5(b).

It is instructive to observe the composition of the standard NTSC

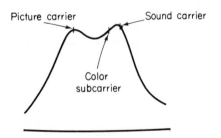

Figure 8–3 Example of a normal frequency–response curve for an RF tuner in a color receiver.

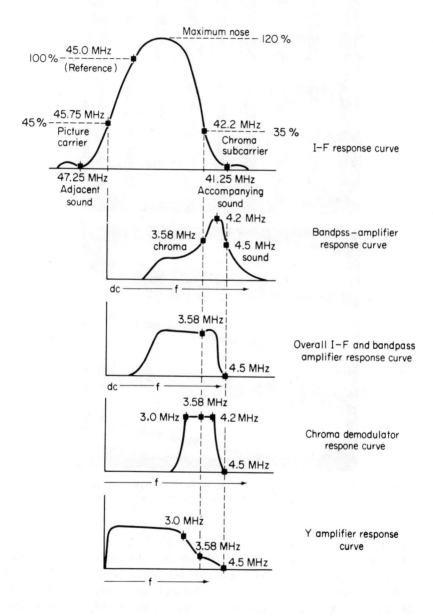

Figure 8–4 Frequency responses for the picture–signal channels in a typical color receiver.

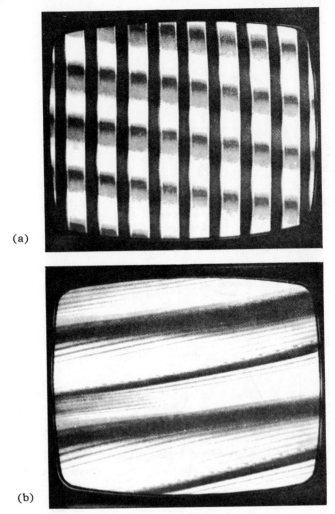

(a)

(b)

Figure 8–5 Picture symptoms resulting from loss of sync lock. (a) Loss of color
sync only; (b) Loss of both color and black–and–white sync.

color–bar signal, as depicted in Fig. 8–6. Note that the Y signal has a
comparatively low video frequency, whereas the chroma signal has a
comparatively high frequency of 3.579545 MHz (rounded off to 3.58
MHz). In turn, it is evident that if the high–frequency response of the
RF tuner or IF amplifier is deficient, the chroma signal will be attenuated, although the Y signal will not be greatly affected. The same ob-
servation applies to the color burst. Figure 8–7 illustrates the appearance
of an NTSC color–bar signal as displayed on the screen of a wide–band

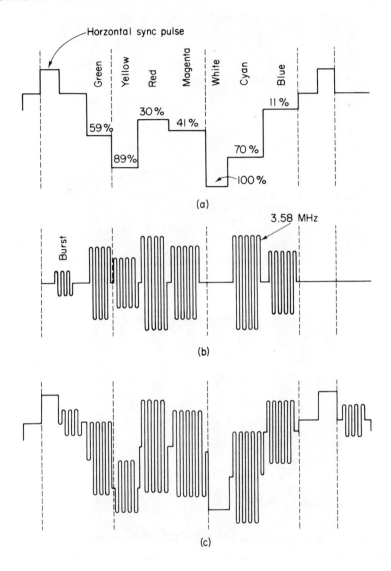

Figure 8–6 Composition of the standard NTSC color–bar signal. (a) Y signal component; (b) Chroma signal component; (c) Saturated color–bar signal.

oscilloscope. This signal produces the color–bar pattern illustrated in Fig. 8–8. An oscilloscope is the most useful signal–tracing instrument in color–TV troubleshooting procedures. A low–capacitance probe is essential to avoid undue circuit loading.

Figure 8–9 illustrates the keyed–rainbow chroma–bar signal that is

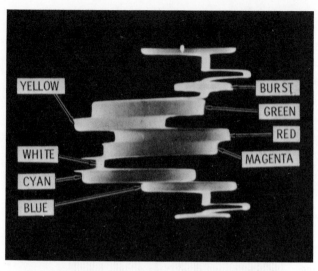

Figure 8–7 Display of an NTSC color–bar signal on the screen of an oscilloscope.

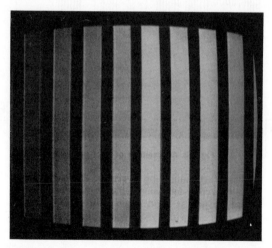

Figure 8–8 Standard NTSC color–bar pattern, displayed on the screen of a color
picture tube.

generally utilized by technicians in color–TV troubleshooting procedures.
This test signal has no Y component, and consists of 11 bursts between
successive horizontal–sync pulses. The burst frequency in a rainbow
signal is 3.563795 MHz (rounded off to 3.56 MHz). A normal keyed–
rainbow chroma–bar pattern displayed on the screen of a color picture
tube is illustrated in Fig. 8–10. If the frequency response of the RF tuner

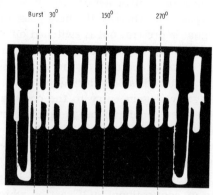

Figure 8–9 Display of a keyed–rainbow chroma–bar signal on the screen of an oscilloscope.

Figure 8–10 A keyed–rainbow chroma–bar pattern, displayed on the screen of a color picture tube.

or **IF** amplifier is deficient, the chroma signal will be attenuated and the bar pattern will appear pale and weak, although the color–intensity control has been advanced to maximum. Note in passing that a rainbow signal is also called a sidelock signal, offset color subcarrier, or linear phase sweep.

8–3 COLOR–KILLER AND ACC TROUBLE SYMPTOMS

With reference to Fig. 8–2, note that the color killer is actuated by the color–burst signal, and controls the operation of the chroma–bandpass amplifier. Two modes of control action are provided. The first mode provides a switching action, whereby the bandpass amplifier is switched on when a color signal is present, but is switched off when a black–and–white signal (or no signal) is present. The second mode provides gain control (automatic chroma control) of the bandpass amplifier. If the bandpass amplifier is not disabled during black–and–white reception, random noise voltages will gain entry to the chroma channel and produce colored interference (confetti) on the picture–tube screen. Therefore, a color–killer level control is provided, whereby the threshold of the electronic switching action can be set for normal control action.

It is helpful to have automatic chroma control (ACC) action in the chroma channel, because the level of the chroma signal tends to vary somewhat in spite of conventional AGC action which is controlled by the black–and–white signal component. Therefore, the color–killer section is arranged to do double duty, and to provide ACC action during color reception. Picture trouble symptoms that point to component defects in this section of the receiver include no color reproduction, confetti during black–and–white reception, color reproduction on strong signals only, intermittent or fluctuating color reproduction, abnormally high color intensity, subnormal color intensity, or rapid changes in color intensity. Note that localization in the color–killer and ACC section is confirmed on the basis of waveform analysis with an oscilloscope and low–capacitance probe.

Figure 8–11 shows the configuration for a typical color–killer, bandpass amplifier, and color–sync arrangement. Note the 250–k color–killer control at the bottom of the diagram. If confetti appears during black–and–white picture reception, the color–killer is probably set too low. On the other hand, if no color is reproduced during color–picture reception, the color–killer is probably set too high. If confetti cannot be eliminated or color cannot be reproduced by adjustment of the color–killer control, the most likely defect is a leaky or open capacitor in the color–killer section. For example, with reference to Fig. 8–11, if the 150–pf capacitor to the base of Q3 is "open," or if the 0.047–μf capacitor to the base of Q4 is "open," the color–killer cannot be activated by a color burst. If the 3300–pf capacitor at the collector of Q9 becomes leaky or "shorted," confetti will be displayed regardless of the setting of the color–killer control.

Open capacitors can usually be pinpointed to best advantage by signal–tracing tests. As an illustration, Fig. 8–12 exemplifies a color–burst display which is normally present at both terminals of coupling capacitors in the burst–amplifier circuit. Leaky and "shorted" capacitors can usually be pinpointed by DC–voltage measurements. In doubtful situations, resistance measurements with a low–power ohmmeter can

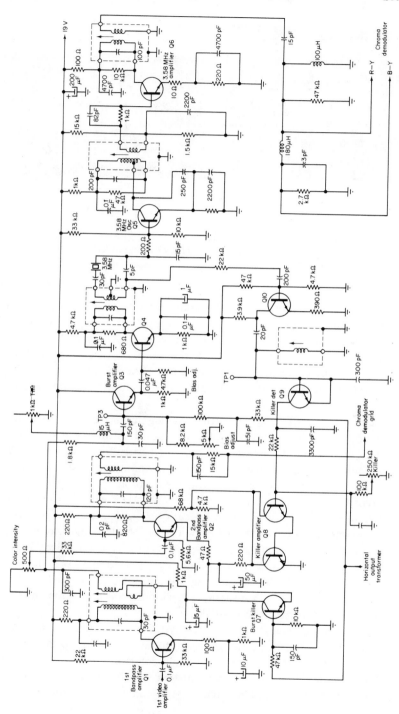

Figure 8–11 Configuration for a typical color-killer, bandpass amplifier, and color-sync arrangement.

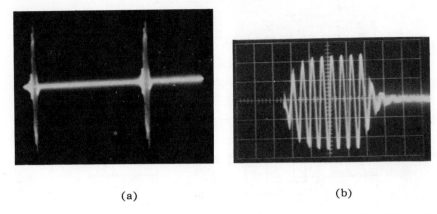

(a)   (b)

Figure 8–12 Displays of a color–burst waveform on the screen of a wide–band
scope. (a) Sweep rate of 15,750 Hz; (b) Expanded waveform (high
sweep rate).

often provide helpful test data. When color reproduction is obtained on
strong incoming signals only, the trouble may be found in either the
black–and–white section or the chroma section of the receiver. Therefore,
the signal waveform at the input of the chroma section should be
checked at the outset. If this signal waveform is normal, as specified by
the receiver service manual, the trouble will be found in the chroma
section.

If color is reproduced only with a strong input signal, check the
adjustment of the color–killer control at the outset. A marginal threshold
adjustment can attenuate the applied signal. Slightly leaky capacitors in
the color–killer section can also cause this symptom. Intermittent or fluctuating color reproduction is often the result of a worn and "noisy"
color–intensity, tint, color–killer, or bias control (Fig. 8–11); that is,
potentiometers become unstable and erratic as a consequence of worn
resistance elements. Intermittents can be caused by defective capacitors,
and may be either mechanical or thermal types. A mechanical intermittent can often be found by tapping suspected capacitors. Thermal
intermittents may be located by heating the pigtails of suspected capacitors with a soldering gun. Transistors occasionally become intermittent,
although the probability of this is small.

Next it is instructive to observe the ACC arrangement associated
with Q1 in Fig. 8–13. Picture trouble symptoms that point to defects in
the ACC section include abnormally high color intensity, subnormal
color intensity, and drifting or intermittent color reproduction. DC–
voltage measurements, supplemented by resistance measurements with

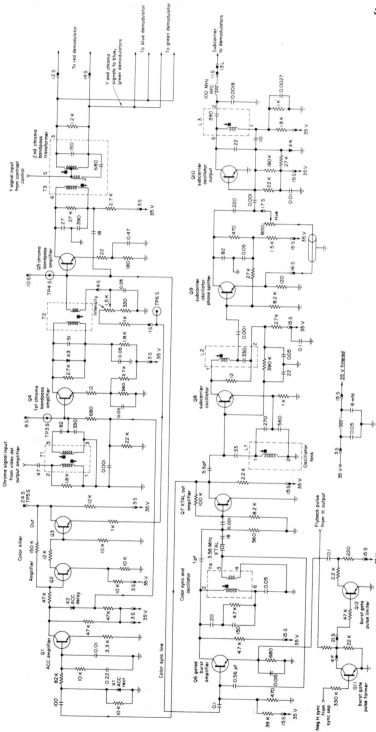

Figure 8–13 Configuration for a typical color–killer and ACC, bandpass amplifier, and color–sync arrangement.

a low–power ohmmeter, are employed to close in on the defective component. Note that an oscilloscope cannot provide test data, because the
ACC section is a DC system. Typical causes of abnormally high color
intensity include leaky capacitors, such as the 100–pf coupling capacitor
in the X1 ACC diode circuit, an "open" ACC delay diode, transistor
with excessive leakage current in the bandpass amplifier, or an off–
value resistor in the ACC section. Note that a defective ACC transistor
must be replaced with one that has closely similar operating characteristics.

Subnormal color intensity can be caused by trouble in the bandpass
amplifier or other chroma sections, in addition to the ACC section. Therefore, with reference to Fig. 8–13, the technician sectionalizes the trouble
area by checking the 3.58–MHz voltage applied to X1. This voltage can
be measured with a TVM or a wide–band oscilloscope. If the driving
voltage is normal, the trouble will probably be found in the ACC section.
Possible causes of subnormal color intensity include a poor front–to–back
ratio in the ACC rectifier X1, leaky capacitors in the base circuit of Q1,
a poor front–to–back ratio in the ACC delay diode X2, or a marginally
defective ACC–amplifier transistor Q1.

Drifting or intermittent color reproduction can be caused by defects
in the antenna, in the signal channels, or in the color picture tube. Accordingly, localization tests are required at the outset. To check the
operating stability of the ACC section, the technician clamps the ACC
output voltage with batteries or with a bias box. Then, if color reproduction is stabilized, he concludes that the trouble will probably be
found in the ACC section. Component defects that cause drifting or
intermittent color reproduction are basically marginal faults in the same
components that cause abnormal or subnormal color intensity. As an
illustration, if a capacitor is leaky, and its leakage resistance tends to
drift, color reproduction can be similarly affected.

8–4 COLOR SYNC TROUBLESHOOTING

Picture symptoms resulting from defects in the color–sync section include loss of color–sync lock (Fig. 8–5), unstable color–sync action,
normal color–sync action on strong signals only, and drifting hues with
normal color–sync lock. Note that the screen pattern illustrated in Fig.
8–5(b) does not necessarily indicate loss of color–sync lock. (The color
burst is gated from the horizontal–sync waveform, as depicted in Fig.
8–14.) In turn, when black–and–white horizontal sync lock is broken,
the gating pulse becomes mistimed, and color–sync lock is lost. There-

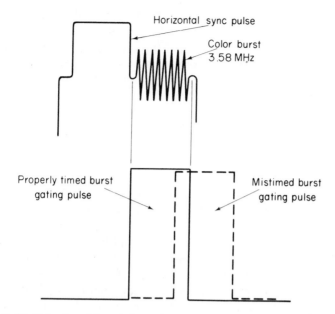

Figure 8–14 Color burst is gated out from the horizontal–sync pulse by a gating pulse.

fore, horizontal–sync action must be restored to normal before color–sync action can be analyzed. Then, if the screen pattern appears as shown in Fig. 8–5(a), it indicates that there is a component defect in the color–sync section.

With reference to Fig. 8–13, the color–sync section comprises Q6, Q7, Q11, and Q12. Note that Q11 and Q12 develop a gating pulse, as depicted in Fig. 8–14. In normal operation, the color burst is gated out of the chroma signal at Q6 as exemplified in Fig. 8–12. Thus, a wide–band scope is the most useful instrument to signal–trace through the color–sync section. In the example of Fig. 8–13, the color burst shock–excites the 3.58–MHz quartz crystal, thereby regenerating the color subcarrier. Output from the ringing crystal is applied to Q8, which operates as a locked (synchronized) oscillator. Three tuned circuits are utilized in the color–sync and subcarrier–oscillator sections, and are resonated at 3.58 MHz.

Statistically, fixed capacitors are the most common troublemakers. Resistors that are sharply pulsed, such as the resistors in the Q11–Q12 circuitry (Fig. 8–13) tend to deteriorate more rapidly than resistors that carry DC or low–frequency sine–wave currents. Note that worn operating controls, such as the hue control in the collector circuit of Q9, are ready suspects when color sync lock is normal, but the reproduced hues tend to drift and vary. Quartz crystals are rarely defective, although this component defect can occur. For example, if a circuit defect causes a quartz crystal to break into violent oscillation, the crystal can become damaged or destroyed. Transistors in the color–sync and color–oscillator sections are ordinarily very long lived, but can be damaged if an associated circuit defect causes excessive current flow.

8–5 CHROMA BANDPASS-AMPLIFIER TROUBLESHOOTING

Picture symptoms resulting from component defects in the chroma-bandpass section include weak hue reproduction or no color, distorted hues, poor color "fit," and/or abnormally intense color reproduction. With reference to Fig. 8–13, the bandpass–amplifier section is comprised of Q4 and Q5. It is evident that a weak–color or no–color symptom, for example, could be caused by a component defect in some section other than the bandpass amplifier. Therefore, a waveform check should be made at the outset with a scope at the input of the bandpass amplifier (test point TP3S to T1). If a waveform is observed as shown in Fig. 8–15, with normal amplitude, it is logical to conclude that the trouble will be found in the bandpass amplifier. Note that the ACC voltage should be clamped in most testing situations.

One common cause of weak or no–color reproduction is a mistuned bandpass–amplifier transformer, which generally results from an associated capacitor defect. For example, a typical bandpass–amplifier frequency–response curve is depicted in Fig. 8–4. If any one of the capacitors associated with T1, T2, or T3 in Fig. 8–13 "opens" or becomes leaky, the transformer will become mistuned, with resulting weak or no–color

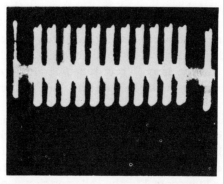

Figure 8–15 Normal waveform at the secondary of the bandpass–amplifier transformer.

reproduction. Therefore, a helpful preliminary analysis can be made by checking the frequency–response curve of the bandpass amplifier with a sweep–and–marker generator and a scope. In the case of older receivers that have been in extensive use, it is good practice to make a routine check of the color–intensity control. A worn and "noisy" control can cause a weak or no–color picture symptom.

Distorted hues, often accompanied by poor color "fit," can result from certain bandpass–amplifier defects. As an illustration, if the 18–pf neutralizing capacitor for Q5 (Fig. 8–13) "opens," this stage becomes regenerative. In turn, its frequency–response curve becomes narrow, and rapid phase shift occurs through the chroma channel. The result is distortion of various hues, and an abnormal time delay is imposed on the chroma signal. This time delay appears in the displayed picture as poor color "fit." Somewhat the same difficulty can occur if a medium–gain bandpass–amplifier transistor is replaced with a very–high–gain transistor. That is, if an exact replacement is impractical, an equivalent replacement is advisable.

Abnormally intense color reproduction is often caused by ACC defects. Accordingly, the ACC line should be clamped at the outset. If the symptom persists, the most likely suspect is a leaky base–coupling capacitor that shifts the base bias. Note that a very leaky or "shorted" emitter–bypass capacitor can cause the same picture symptom. Both of these component faults can be pinpointed by means of DC–voltage measurements. A defective color–intensity control can be responsible for excessively intense color, as can an incorrect type of replacement transistor. Off–value resistors sometimes also cause bias shift, although these are statistically much less probable causes than defective capacitors or worn controls.

8-6 CHROMA DEMODULATOR AND MATRIX TROUBLESHOOTING

Picture trouble symptoms caused by component defects in the chroma–demodulator section often involve phase errors. As depicted in Fig. 8–16, each hue corresponds to a specific chroma phase angle. Typical symptoms caused by chroma–demodulator or matrix malfunctions are incorrect relative hues, all hues incorrect, some hues missing, or incorrect intensities of particular hues. Although various chroma–demodulator and matrix arrangements are employed, the end result is always the same, and the troubleshooting approaches are basically the same. One of the most widely used arrangements is depicted in Fig. 8–17. This is an

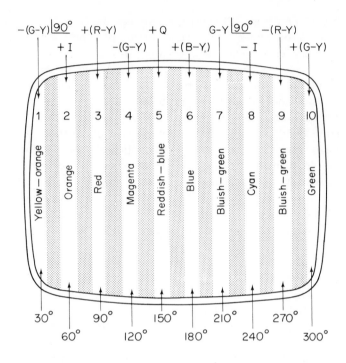

Figure 8–16 Each hue corresponds to a certain chroma phase angle with respect to burst.

example of **X** and **Z** demodulation at 120°. Outputs from the two demodulators are matrixed, and red, green, and blue chroma signals are applied to the respective grids of the color picture tube.

When the relative hues are incorrect, it is most likely that the 3.58–MHz subcarrier voltage is being injected into one of the chroma demodulators with incorrect phase. As an illustration, if C1 (Fig. 8–17) becomes leaky or "open," the subcarrier will be injected into the emitter circuit of Q2 with a phase error. Demodulation phase errors are checked to best advantage by means of a vectorgram display, as exemplified in Fig. 8–18. A vectorgram is produced by feeding the demodulator (or matrix) outputs to the vertical and horizontal inputs of an oscilloscope. For example, with reference to Fig. 8–17, the CRT red–grid output would be applied to the vertical–input terminals of the scope, and the CRT blue–grid output would be applied to the horizontal–input terminals of the scope. The demodulation phase is given by the ellipticity of the vectorgram, as for standard Lissajous patterns.

When all hues are incorrect, it is indicated that the 3.58–MHz oscillator voltage (Fig. 8–17) has been previously substantially shifted in phase, so that the hue control is out of range. For example, with reference to Fig. 8–13, the technician would suspect that a capacitor defect would be found in the subcarrier–oscillator output stage, or in the subcarrier–oscillator phase–splitter stage. In this situation, a systematic check of the capacitors is required. Note that the possibility of a worn and "noisy" hue control should not be overlooked. In some cases, components change values slightly due to aging, and a basic phase readjustment is all that is required. For example, the slug in L3 would be adjusted in this situation, to bring the hue control within its normal range.

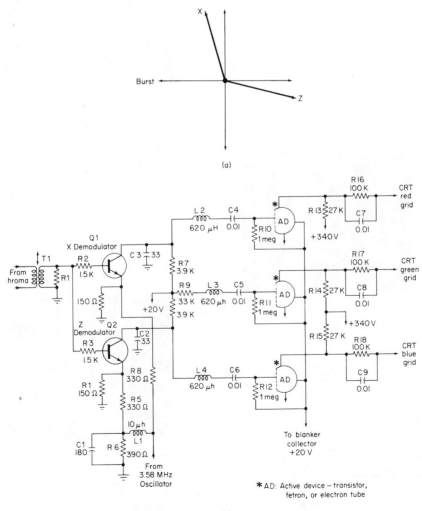

(a)

(b)

* AD: Active device — transistor, fetron, or electron tube

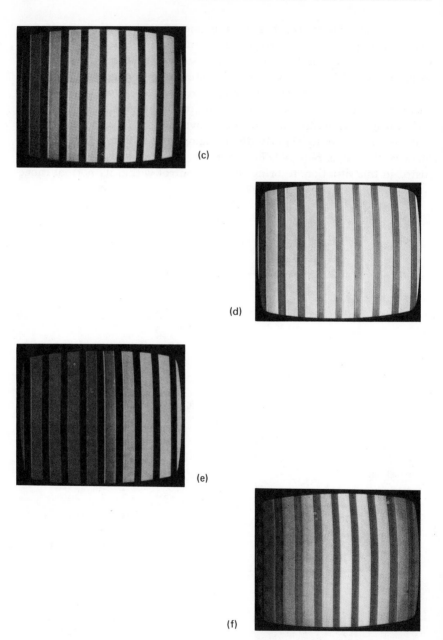

(c)

(d)

(e)

(f)

Figure 8–17 Example of **XZ** chroma demodulation with R–Y, B–Y, and G–Y matrixing. (a) Typical **X** and **Z** chroma phases; (b) **XZ** demodulator and matrix arrangement; (c) Normal keyed–rainbow display; (d) Display with **X** demodulator inoperative; (e) Display with **Z** demodulator inoperative; (f) Display with G–Y matrix inoperative.

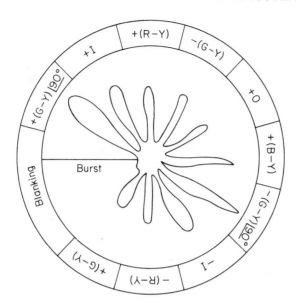

Figure 8–18 Demodulation phase and demodulated phases displayed by a keyed–rainbow vectorgram.

When some hues are missing, it is most likely that one of the demodulator or matrix stages is inoperative. As an illustration, if the X demodulator becomes inoperative, a keyed–rainbow pattern appears as in Fig. 8–17(d). Or if the Z demodulator becomes inoperative, a keyed–rainbow pattern appears as in Fig. 8–17(e). Again, if the G–Y (green) matrix becomes inoperative, a keyed–rainbow pattern appears as in Fig. 8–17(f). In most cases, the culprit responsible for signal blocking is an "open" or "shorted" capacitor. However, a demodulator or matrix transistor will occasionally become defective, particularly if a "shorted" capacitor has caused excessive current flow through the transistor. Other occasional culprits are cold–soldered connections, cracked PC boards, and "shorts" due to solder splashes.

Various types of vectorscopes are commercially available, and they provide considerable assistance in troubleshooting a number of faults in chroma demodulator and matrix circuits. Figure 8–19 depicts the configuration of a basic vectorscope. Note that it is essentially a cathode–ray tube with supply voltages, operating controls, and capacitive coupling to the vertical and horizontal deflection plates. This is a low–sensitivity instrument, and can be driven only by the high–level chroma signals at the grids of the color picture tube. More elaborate vectorscopes, such as noted in Chapter 2, have vertical and horizontal am-

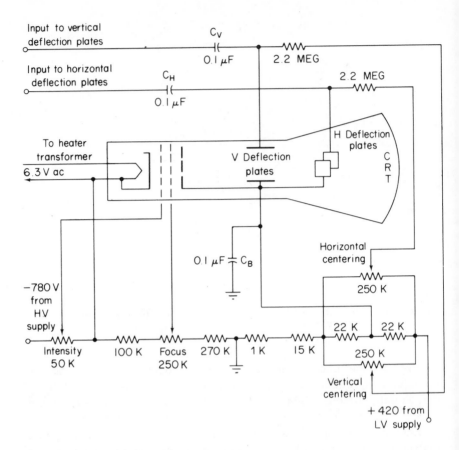

Figure 8–19 Basic vectorscope arrangement.

plifiers with negligible distortion through the chroma frequency range. These are high–sensitivity instruments which can be used effectively to troubleshoot low–level chroma circuits.

We will find that vectorgrams are based on Lissajous patterns. Thus, if we apply sine–wave voltages of the same amplitude and frequency, but with a phase difference of 90° to the vertical and horizontal deflection plates, respectively, we observe a circular pattern as depicted in Fig. 8–20. In practice, most vectorgrams have an elliptical outline. Figure 8–21 shows typical Lissajous patterns and their corresponding phase angles. Note that if the vertical and horizontal deflection voltages have the same amplitude, any ellipse that is displayed will lean 45° with respect to the vertical and horizontal axes. However, if the vertical–deflection voltage is greater or less than the horizontal–deflection voltage,

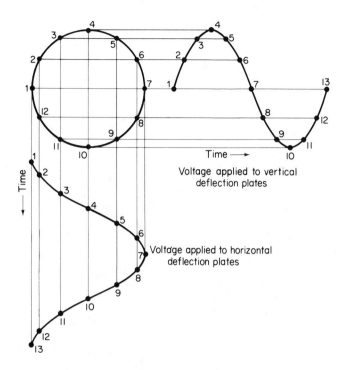

Figure 8–20 Formation of a circular Lissajous pattern.

we observe patterns such as those depicted in Fig. 8–22. These ellipses do not lean with respect to the vertical and horizontal axes.

Next, if a 120° ellipse is formed as shown in Fig. 8–21, and the amplitude of the vertical signal is increased, the ellipse will be displayed at a greater height and will lean less than 45° with respect to the vertical axis. These are two general methods of determining the phase angle corresponding to an elliptical Lissajous pattern. If the vectorscope provides adjustable vertical and horizontal gain, the pattern is adjusted to be tangental to a square as exemplified in Fig. 8–21. In turn, the phase angle can be approximated by comparison with the examples given in the diagram. Although this method is not highly accurate, it is adequate for chroma–circuit troubleshooting tests.

A more accurate method of measuring phase angles is shown in Fig. 8–23. The pattern is centered on the screen, and the distances a and b are measured. In turn, the fraction a/b is equal to the sine of the phase angle, and this angle can be found in any trigonometry book from a table of sines. The ratio a/b remains unchanged regardless of

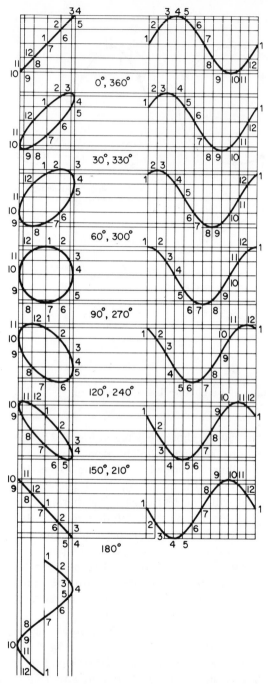

Figure 8–21 Lissajous figures formed by sine waves with various phase angles.

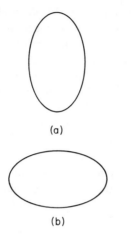

(a)

(b)

Figure 8–22 Lissajous figures formed by in–phase sine waves with differing amplitudes. (a) Vertical signal greater than horizontal signal; (b) Horizontal signal greater than vertical signal.

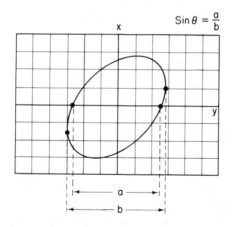

Figure 8–23 Generalized phase–angle measurement.

the amplitudes of the vertical and horizontal signals. Therefore, this method is used when the vectorscope provides no control of vertical and horizontal gain. The significance of phase measurements to chroma demodulator and matrix circuit action is explained subsequently.

The simplest type of color test signal is the unkeyed–rainbow signal. It consists of a 3.56395–MHz sine wave, which is 15,750 Hz less than the color–subcarrier frequency of 3.579545 MHz. Because of this frequency difference, an unkeyed–rainbow signal shifts progressively out of phase with the receiver's color–subcarrier oscillator, and these two signals pass through an in–phase state at the start of each forward-scanning interval. Figure 8–24 depicts the basic vectorscope test setup and typical vectorgrams. A circular pattern is displayed when demodulation occurs along the R–Y and B–Y axes (90° demodulation angle), and the amplitudes of the R–Y and B–Y signals are equal. The "pie cut" is caused by horizontal–blanking action in the receiver. Again, if the demodulation angle is not 90°, an ellipse with a "pie cut" is displayed.

Now let us note some typical trouble indications and their causes. Thus, a leaky capacitor in the subcarrier phase–shifting network will cause the chroma demodulation angle to shift from a normal value of

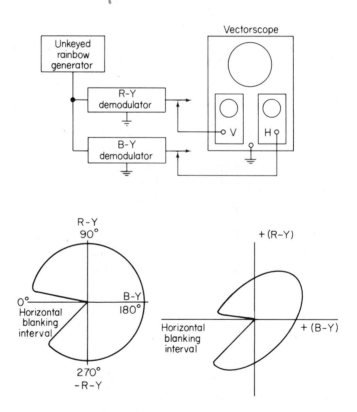

Figure 8–24 Basic vectorscope test setup using an unkeyed–rainbow signal. Circular pattern results when vertical and horizontal signals are equal and chroma circuits are operating normally. A tilted ellipse results when chroma demodulation angle is not 90 degrees.

90°. In such a case, an elliptical vectorgram is displayed instead of a circular vectorgram, as was shown in Fig. 8–24. As a note of caution, different receivers have different normal demodulation phase angles. Therefore, it is essential to consult the receiver service data before concluding that an elliptical vectorgram display is necessarily a trouble indication. This topic is developed in greater detail subsequently.

We will find that various forms of nonlinear distortion may occur, due to defective components in chroma circuitry. Thus, off–value resistors or leaky capacitors can cause incorrect bias voltages, with consequent compression or clipping of the chroma signal. Fig. 8–25 depicts a vectorgram that results from positive–peak clipping of the R–Y signal. This type of distortion produces desaturated and distorted hues, which are particularly evident as incorrect flesh tones in reproduced images. In this situation, the distorted vectorgram identifies the basic trouble and points to the circuit area that is at fault.

Next let us consider the keyed–rainbow signal depicted in Fig. 8–26.

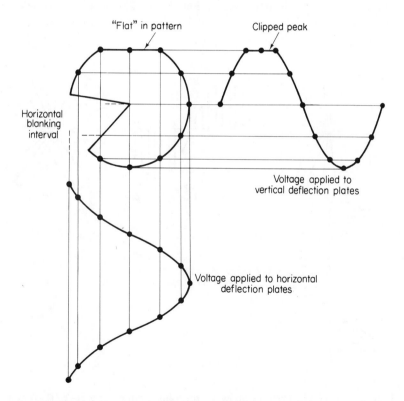

Figure 8–25 Vectorgram produced by positive–peak clipping of the R–Y signal.

Figure 8–26 Waveform of the keyed–rainbow signal.

It has the same basic characteristic as an unkeyed–rainbow signal. However, the keyed signal is more informative because its sequence of bursts serves to identify 10 chroma phases. Using the same test setup as in Fig. 8–24, but with a keyed–rainbow generator, the ideal vectorgram that results is depicted in Fig. 8–27. The tops of the R–Y and B–Y bar signals fall along a 15,750–Hz sine wave in this ideal waveform,

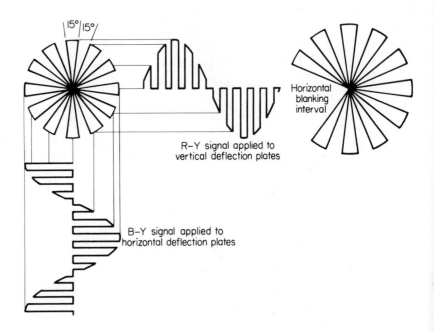

Figure 8–27 Ideal keyed–rainbow vectorgram produced by R–Y/B–Y demodulators. Note that two petals are normally blanked.

and the tops of the vectorgram petals fall along the circumference of a circle.

However, in actual practice, the tops of the R–Y and B–Y bar signals do not have sharp corners, but are rounded, as depicted in Fig. 8–28. In turn, the tops of the vectorgram petals are rounded correspondingly. Corner rounding of chroma–bar waveforms results from limited high–frequency response, as in square–wave or pulse test responses. Note also that the R–Y and B–Y bar signals in Fig. 8–28 are idealized in that they are shown with zero rise time. In actual practice, the rise time is slowed down to an extent determined by the bandwidth of the chroma circuits. In turn, the bar signals become "feathered" to some extent, just as a horizontal sync pulse becomes "feathered" in passage through an integrating circuit.

An important type of vectorgram pattern is produced when the bar signals happen to have half–sine waveforms. This is a specific form of

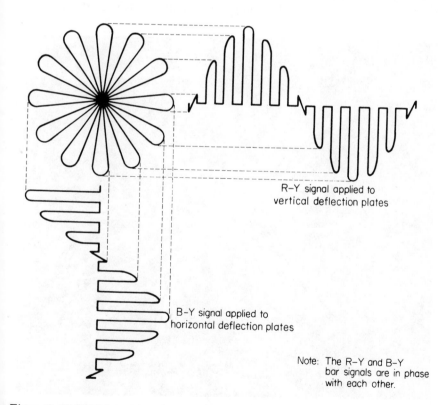

R–Y signal applied to vertical deflection plates

B–Y signal applied to horizontal deflection plates

Note: The R–Y and B–Y bar signals are in phase with each other.

Figure 8–28 Vectogram petals are rounded in accordance with the chroma–circuit bandwidth.

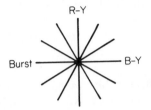

Figure 8–29 Vectorgram has straight–line petals if the bar signals have a half–sine waveform.

"feathering" which results in a vectorgram that consists of single straight lines, as depicted in Fig. 8–29. This form of display may be partially realized in practice, as exemplified in Fig. 8–30. We observe that three of the vectorgram petals approximate straight lines, although the remaining petals have appreciable widths. Another practical point to be observed in Fig. 8–30 is that only nine petals appear in the pattern,

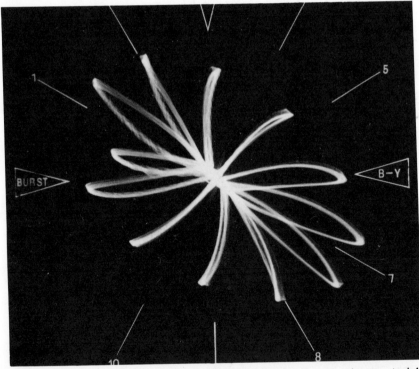

Figure 8–30 A vectorgram in which three of the petals approximate straight lines (*Courtesy of* Radio-Electronics).

whereas an ideal pattern has 10 petals. The "lost petal" is the result of a comparatively long horizontal–blanking interval, or slow horizontal fly-back. We will also note that the horizontal–blanking interval in Fig. 8–30 appears at the upper right–hand portion of the vectorgram. This aspect results from the fact that the cathodes of the color picture tube were driven in this example.

Next, consider the situation in which one bar signal is slightly dis-placed in time (lags or leads the corresponding bar signal). Thus, if the bar signals have half–sine waveforms, and the B–Y signal lags the R–Y signal slightly, the vectorgram that results has elliptical petals, as depicted in Fig. 8–31. Again, let us consider the situation in which the R–Y and B–Y signals are exactly in phase, but with the R–Y bars slightly wider than the B–Y bars. In such a case, the half–sine wave-forms do not form straight lines, but petals with an appreciable width. The result is somewhat similar to the wider petals appearing in the pattern of Fig. 8–30.

Note in Fig. 8–30 that the petals extend down to the center of the vectorgram, where they form a bright spot. It follows that the R–Y and B–Y chroma channels have adequate bandwidth. On the other hand, if the chroma channels have inadequate bandwidth, the center of the re-sulting vectorgram becomes distorted. If there is inadequate high–fre-quency response, the center of the vectogram becomes an "open circle," as depicted in Fig. 8–32. This is the result of baseline curvature in the R–Y and B–Y signals which is produced by inadequate high–frequency response of the chroma channels.

It is instructive to observe the chief distortions that appear in the vectorgram shown in Fig. 8–33. We observe nonlinear distortion, in-dicated by the eccentricity of the vectorgram. Thus, the +Q petal is quite a bit longer than the +I petal. We also observe overload distor-tion in the tops of the −I, −(R–Y), and +(G–Y) petals in particular. Note, however, that the +(B–Y) petal shows little or no overload dis-

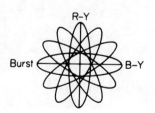

Figure 8–31 Wide petals in a vectorgram caused by slightly out–of–phase R–Y and B–Y signals.

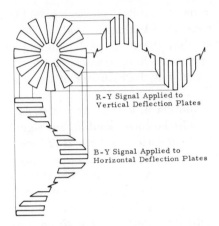

Figure 8–32 Inadequate high–frequency chroma response causes an "open circle" in the vectorgram center.

Figure 8–33 A distorted vectorgram pattern (*Courtesy of* Radio-Electronics).

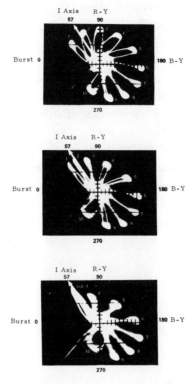

Figure 8–34 Systematic changes in a vectorgram pattern produced by advancing the automatic tint control.

tortion. Nonlinear distortion is usually caused by incorrect base or grid bias. Leaky capacitors are the most common culprits. Finally, as a practical precaution, it is important to switch off the automatic tint control of a receiver before analyzing a vectorgram. The automatic tint control produces systematic changes in a vectorgram pattern, as exemplified in Fig. 8–34. Of course, if a vectorgram is otherwise normal, these systematic changes produced by advancing the automatic tint control serve to verify that this control is functioning normally.

QUESTIONS

1. How is the color killer activated?
2. How should the frequency response of the RF tuner and the IF strip be checked in a color television receiver?

3. What is the relationship between the frequencies of the Y signal frequency and the chroma frequency?

4. What are the names for a rainbow signal?

5. What are the symptoms if the bandpass amplifier is not disabled during black–and–white reception?

6. What is the first check that should be made if color is reproduced only on strong signals?

7. What are some of the causes of intermittent or fluctuating color reproduction?

8. How may thermal intermittents of capacitors be found?

9. What are the picture symptoms that point to the ACC section?

10. Where can problems that cause drifting or intermittent color reproduction be located?

9

AMATEUR, CB, AND MOBILE RADIO TROUBLESHOOTING

9-1 GENERAL CONSIDERATIONS

Troubleshooting amateur radio receivers involves the same basic principles that apply in servicing of AM and FM broadcast receivers. However, some specialized functions are employed in amateur receivers, such as provision for single–sideband reception, double conversion, and quartz crystal filters. Troubleshooting amateur radio transmitters requires test methods and analytical procedures suitable for signal generating equipment, instead of signal processing equipment. Citizens-band receivers differ from broadcast and amateur radio receivers in the use of crystal local oscillators. Dual conversion is also utilized in some CB receivers. CB transmitters are basically similar to low–power amateur transmitters. Mobile receivers and transmitters are generally characterized by narrow–band FM design. Since an FCC license is required for any major adjustment or servicing of radio transmitters, this is a somewhat specialized field.

A typical trouble symptom in amateur radio–receiver operation is "birdies" (heterodyne squeals or chirps) caused by oscillation in the high–frequency amplifier or mixer stages. If strong oscillation occurs, there may be weak or no reception. Frequency drift is another typical trouble symptom, requiring periodic resetting of the tuning dial. Motor-boating may occur, as in broadcast receivers. Misalignment causes weakened output, often accompanied by poor selectivity and interference. Alignment is particularly critical in elaborate high–performance receivers that utilize narrow passbands with sharply defined frequency limits.

Troubleshooting amateur radio transmitters involves problems of incomplete neutralization, modulation nonlinearity, chirp, drift, parasitic oscillation, interference to television or radio receivers, and weak or no output. Note that routine maintenance procedures such as tuning up and antenna loading are not included in the troubleshooting category. On the other hand, if a transmitter cannot be tuned up properly, or if an excessively high standing–wave ratio cannot be reduced, a troubleshooting problem presents itself. Since lethal power supplies are used in all transmitters that have appreciable power output, fatalities can be avoided only by adequate precautionary measures. For example, even if a transmitter is turned off and disconnected from the power line, an unwary troubleshooter can be killed by the stored charge in the filter capacitors. Therefore, a rigid rule to be observed in all cases is to discharge filter capacitors before touching any transmitter circuitry.

9–2 BASIC RECEIVER TROUBLESHOOTING

Figure 9–1 shows an example of a simple direct–conversion amateur radio receiver. A tuning range from 3.5 to 4 MHz is provided. This is essentially a single–sideband and CW receiver, although AM signals can also be reproduced by tuning the beat–frequency oscillator to zero beat. Observe that incoming signals are applied directly to the integrated–circuit detector U1, where they are beat against the BFO output from Q2. In turn, the demodulated output is amplified by Q1 and the IC stage U2. A carrier for ssb reception and a beat note for CW reception are provided by the BFO. Oscillator stability is optimized by zener diode CR1 which regulates the supply voltage to Q2. Effective selectivity is improved by passing the audio signal through a 2.5–kHz bandpass filter. Polarity–guarding diode CR2 prevents damage to the receiver in case the battery might be accidentally connected in wrong polarity. CR3 and CR4 protect the receiver from damaging overload by nearby transmitters.

Amateur radio–receiver diagrams rarely specify normal DC–voltage values throughout the circuits. Therefore, measured voltage values must be evaluated on the basis of experience. If a similar receiver in normal

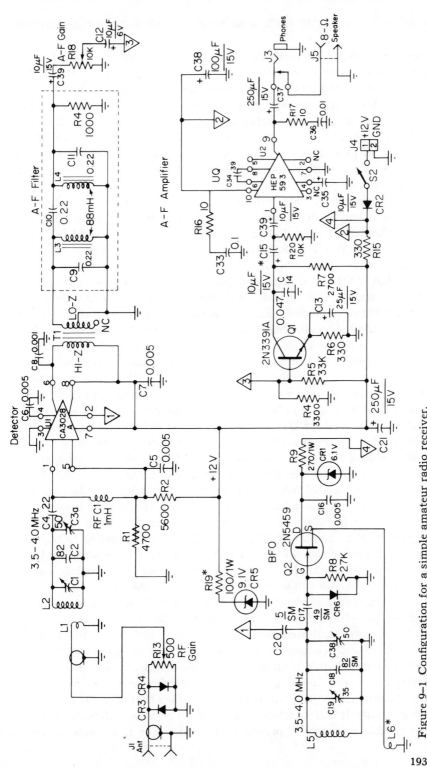

Figure 9-1 Configuration for a simple amateur radio receiver.

193

working condition is available, measured voltage values in the two receivers can be compared. Whenever a receiver is "dead," the battery supply voltage should be checked first. With reference to Fig. 9–1, check next at R15. Zero voltage at this point indicates that the polarity-guarding diode CR2 is probably open–circuited. However, if CR2 is not defective, measure the voltage drops across CR1 and CR5. Zero voltage at either of these points indicates that the associated diode is probably short–circuited. Note that a short–circuited capacitor such as C16 may also cause a zero–volt reading across CR1.

If transistor Q1 (Fig. 9–1) is suspected of being defective, it can be checked in–circuit by means of turn–on and turn–off tests. To determine whether Q2 is oscillating, its output should be checked by tuning in on another receiver along the 8–meter band. A test lead is connected to the antenna–input terminal of the other receiver, and placed near Q2. Note that if Q2 is not oscillating, the receiver will seldom appear completely "dead"; that is, any AM transmission in the 3.5–to–4–MHz range will still be audible. Signal–tracing or signal–substitution tests are basic in trouble localization procedures, and will sometimes pinpoint a defective component. As an illustration, if normal response is obtained when a signal is injected at the input of U1, but no response occurs when the signal is injected at J1, either CR3 or CR4 is short–circuited. Again, if normal response is obtained when a signal is injected at the output of U1 (Fig. 9–1), but no response occurs when the signal is injected at the input of U1, the IC is probably defective. If a capacitor is suspected of being open–circuited, it can be tested by "bridging" it with a known good capacitor. Note that C7 and C8 are cleared from suspicion of short–circuits in the foregoing procedure, because the test showed that an output signal from the IC proceeds through the audio amplifier. In the event that a coupling capacitor is open–circuited, no response will be obtained from signal voltages injected at the input end of the capacitor. For example, if C12, C15, C37, or C39 happens to be open–circuited, the receiver will be "dead" with respect to signals injected at the input end of the defective capacitor. An open–circuited decoupling or bypass capacitor will weaken, distort, or "kill" the signal.

Pin contacts of transistors or ICs in sockets are sometimes faulty. Cold–soldered connections can be deceptive, because the circuit appears to be all right until a continuity test is made across the defective connection. Volume controls tend to become noisy or open–circuited after extensive use. Switch contacts can become defective, particularly after a receiver has been in storage for a long time. Occasionally, the foregoing defects produce intermittent trouble symptoms that occur at unpredictable intervals. It is impractical to close in on an intermittent defect unless the trouble symptom is present at the time. For this reason, intermittents may consume much more troubleshooting time than a catastrophic failure of a component.

9–3 SUPERHETERODYNE AMATEUR RECEIVER SERVICING

Two basic superheterodyne arrangements used in amateur radio receivers are shown in Fig. 9–2. Note that a double–conversion receiver provides improved selectivity (better image rejection). Since the first mixer provides a high IF frequency, such as 5 MHz, image interference is minimized. In turn, the second mixer provides a lower IF frequency, such as 50 kHz, and most of the gain is contributed by the following 50–kHz IF amplifier. Although the circuitry of a double–conversion receiver is somewhat elaborate, servicing procedures are essentially the same as for a conventional arrangement. Figure 9–3 (a, b, and c) depicts a configura-

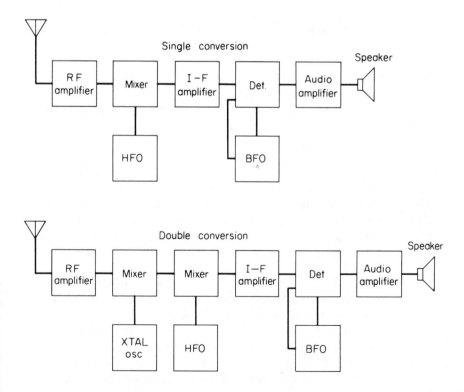

Figure 9–2 Two basic superheterodyne arrangements.

tion for a single–conversion amateur superheterodyne receiver. A 455-kHz mechanical filter is utilized in the IF section to limit the circuit band-width for code reception. The filter is switched out for phone reception. Typical trouble symptoms and probable causes include the following:

Birdies. Chirps that occur as the tuning dial is turned through its range commonly cause this symptom. Defective bypass capacitors, such as C10 and C11 in Fig. 9–3(a), or a poor ground connection, such as R1 to ground or to C5, may result in oscillation and chirps. Violent oscillation in the input section will block the signal entirely. In case of doubt, it is helpful to regard the receiver as a possible transmitter, and to determine if its oscillation frequency can be tuned in on another nearby amateur receiver. This test will quickly confirm or eliminate the suspicion of front–end oscillation.

Note that if oscillation occurs in the IF section, its frequency is not affected by the setting of the tuning dial. When the BFO is switched on, a continuous squeal may be heard. Or, if the oscillation is violent, there will be no audible output. With reference to Fig. 9–3(b), a useful quick check for IF oscillation is to switch off the BFO and measure the output voltage from X1. If a substantial DC–voltage output is found, it is concluded that the IF amplifier is oscillating. Open bypass or decoupling capacitors are common culprits. As an illustration, if C8 becomes open-circuited, the IF amplifier is likely to become unstable. In this situation, oscillation becomes more violent as the internal impedance of the source voltage increases. That is, if a battery source is used, trouble symptoms become more pronounced as the battery ages.

Mushy Hiss. If a mushy hiss occurs instead of chirps as the tuning dial is turned through its range, it is probable that the local oscillator is blocking at an audible rate, in addition to generating a high–frequency output. A common cause of this condition is a gate–leak resistor that has increased greatly in value or has become open–circuited, such as R4 in Fig. 9–3(a). Another possibility is a defective oscillator transistor.

Frequency Drift. A changing beat note during CW reception is caused by frequency drift, either in the high–frequency oscillator or in the beat–frequency oscillator. To sectionalize the trouble, a signal generator can be used to inject an IF signal into the IF amplifier. Then, if the beat note is steady, the trouble will be found in the high–frequency oscillator. A defective capacitor or a faulty transistor is the most likely cause. The possibility of a poor connection should not be overlooked. In home–built receivers, frequency drift may occur because of poor design practices. Line–voltage changes can cause frequency drift, and it is good practice to employ supply–voltage regulation for oscillators. Voltage regulation also ensures that motorboating will not occur during phone reception with AVC operation. When frequency drift occurs in

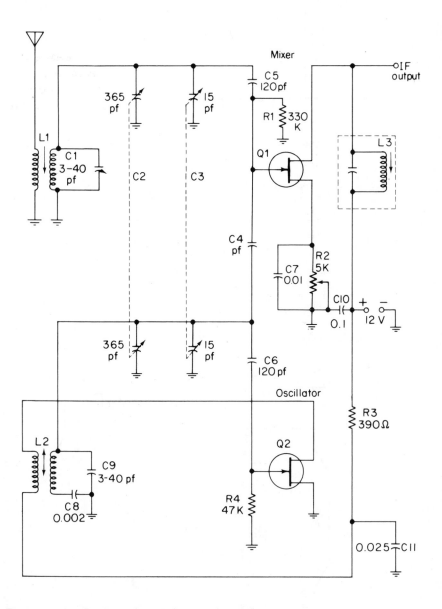

Figure 9–3(a) Input section of the superheterodyne.

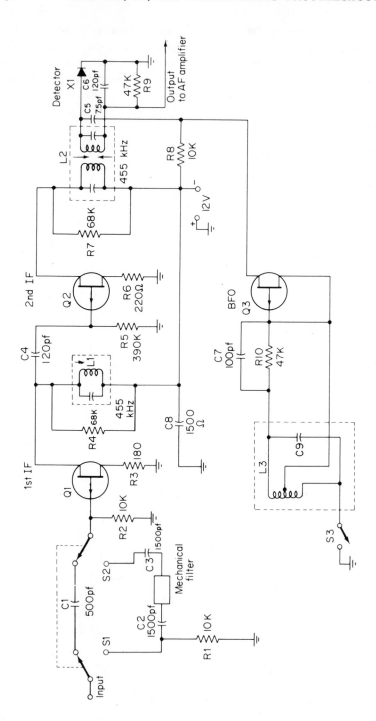

Figure 9-3(b) IF section of the superheterodyne.

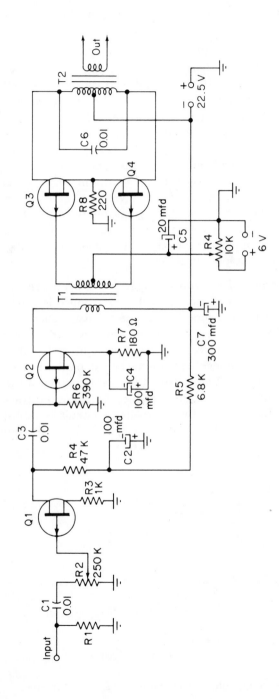

Figure 9–3(c) Audio section of the superheterodyne.

receivers with regulated supply voltages, the regulating devices should be checked.

Weak Output. Weak output is caused by most of the component defects that stop signal passage completely. However, the failure is marginal when the signal is weakened. With reference to Fig. 9–3, such marginal failures can occur in any of the three principal sections of the receiver. Therefore, it is generally necessary to make a sectionalization test at the outset. If the antenna–input terminal is energized by an AM signal generator, the modulation envelope can be traced through the signal channel with an oscilloscope and low–capacitance probe. Stage gains for a typical receiver are shown in Fig. 9–4. However, receiver service data seldom specify gain values. Therefore, it is necessary for the technician to check stage gains on the basis of experience, or by comparison with a receiver that is in good operating condition. Note that weak reception can be caused by tuned–circuit misalignment as well as by marginal component defects.

Alignment Procedures. Alignment may be required after component replacement in the signal channels, and it is good practice to check the alignment of a superheterodyne receiver at least once a year. Commercial receivers are generally provided with detailed instruction manuals that explain correct alignment procedures. Home–built receivers are aligned in accordance with the technical proficiency of the constructor. With reference to Fig. 9–3(a), the tuned circuits in the input section of a superheterodyne should be aligned for optimum tracking. This means that the LC ratio of the mixer–input circuit is adjusted to provide a tuning characteristic that is properly related to the tuning characteristic of the local oscillator.

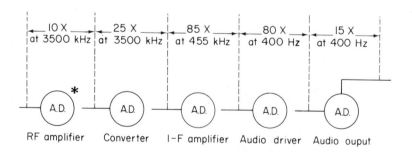

Figure 9–4 Stage voltage–gain values for a typical receiver.

To track the input configuration depicted in Fig. 9-3(a), the tuning capacitors are fully meshed, and the antenna–input circuit is energized by an AM signal generator set to the low–frequency edge of the band. Trimmer capacitors C1 and C9 are approximately adjusted to minimum capacitance. Then the cores in L1 and L2 are tuned to provide maximum output. Some technicians connect an output meter at the output of the audio amplifier to indicate the output level, and others make an approximate judgment on the basis of the sound–output level. The tuning capacitors are then fully unmeshed, and the signal generator is set to the high–frequency edge of the band. Trimmer capacitors C1 and C9 are then adjusted to provide maximum output. For maximum accuracy of alignment, the tracking procedure may be repeated.

IF amplifiers are aligned by peaking each tuned circuit to the IF frequency. With reference to Fig. 9-3(b), the mechanical filter is first switched out, and the cores in L1 and L2 are adjusted to provide maximum output when a 455–kHz generator signal is applied to the input terminal. Then the mechanical filter is switched into the circuit and the generator is retuned slightly, if necessary, for maximum output. Finally, the cores in L1 and L2 are touched up to provide maximum output.

9-4 AMATEUR TRANSMITTER TROUBLESHOOTING

A typical 50–MHz, 40–watt CW transmitter configuration is shown in Fig. 9-5. This arrangement employs a voltage standing–wave–ratio (VSWR) bridge circuit to maintain a safe dissipation level in the output stage under all conditions of antenna mismatch. Higher–powered amateur radio transmitters generally employ tubes. However, the basic troubleshooting principles are the same. Among common trouble symptoms in transmitters, the following conditions may be noted:

Parasitic Oscillation. Either low–frequency or high–frequency parasitic oscillation may be encountered. High–frequency parasitics cause interference, reduce transmitter efficiency, and often overload one or more stages. Low–frequency parasitics cause interference, although the transmitter may seem to be operating properly. This type of parasitic oscillation occurs chiefly in solid–state circuitry. The most likely suspect is a transistor with shunt feed. In this situation, it is advisable to change over to series feed, as depicted in Fig. 9-6. In addition, decoupling of power–supply leads may need to be supplemented by a second capacitor that has a large value and a correspondingly low reactance at the low parasitic frequency, as shown in Fig. 9-7.

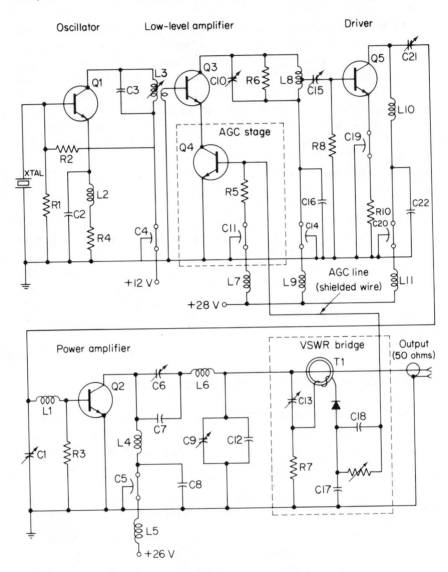

Figure 9–5 Configuration of a 50–MHz, 40-watt CW radio transmitter.

Parasitic oscillation can be checked by means of radio receivers operated in the vicinity of the transmitter under key–up conditions. A TV receiver may also show the presence of parasitic radiation, or a TVM with an RF probe can be used to check in the vicinity of the tuned circuits in a transmitter to localize the trouble. Some amateur

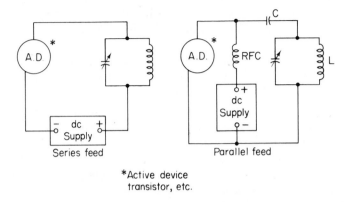

Series feed   Parallel feed

*Active device
transistor, etc.

Figure 9–6 Basic series and parallel feed configurations.

operators use a neon bulb, although this is a less sensitive indicator. One terminal of the neon bulb is touched to a tank–circuit lead to make the test.

Chirp and Drift. A common trouble in home–designed and –constructed transmitters is keying chirp. This is a form of spurious frequency modulation that is usually caused by oscillator load variation or by inadequate power–supply regulation. A keyed or amplitude–modulated oscillator is almost certain to develop chirp to some degree. The

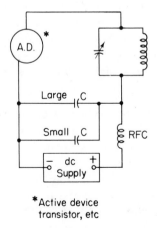

*Active device
transistor, etc

Figure 9–7 Two decoupling capacitors assist in suppression of low–frequency
parasitics.

most practical solution is to provide one or two buffer stages between the oscillator and the driven stage. Using two buffer stages will practically eliminate any load variation on the oscillator when the driven stage is keyed. If there is residual chirp due to oscillator power–supply fluctuation, the oscillator source voltage should be regulated.

Frequency drift inevitably occurs when a transmitter is first turned on from a cold state. However, in a normal operating condition, the radiated frequency quickly stabilizes. When a component defect or poor design causes the frequency to change slowly (or sometimes rapidly) in the key–down condition, systematic checks are required to pinpoint the difficulty. One of the most common causes of drift is inadequate ventilation of frequency–determining devices and circuits (a blower may have failed, or normal circulation of air may have become impeded in some manner). Again, the crystal oscillator may be misadjusted, causing the crystal to gradually heat up abnormally. Parasitic oscillation may occur in a crystal–oscillator arrangement; a small series resistor or a ferrite bead near the base terminal of the oscillator transistor will usually suppress the parasitic.

Neutralization. No transmitter can operate normally unless it is properly neutralized. Neutralizing procedures are usually straightforward, but if there is any difficulty there is usually a parasitic that must be located and suppressed. It may be sufficient to connect a 50–ohm resistor, an RF choke, or both at the grid and plate terminals of the socket, as shown in Fig. 9–8. Approximately six turns of wire may be wound around a 50–ohm resistor. Parasitic oscillation can be detected as whistles in a nearby radio receiver, or as self–developed grid–bias voltage in circuits similar to Fig. 9–8. A TVM with an RF probe will indicate the presence of a parasitic if the probe is brought into the vicinity of the oscillating circuit. If the amplitude of parasitic oscillation is fairly high, a neon lamp will glow when one terminal is touched to the parasitic circuit. This can be a highly dangerous procedure for the inexperienced technician.

Lightning Damage. A lightning strike on an unprotected antenna can be a puzzling cause of transmitter malfunction or failure. Lightning damage is very capricious, and only a systematic checkout can determine the extent of destruction. Wires may be open–circuited or melted together and short–circuited, insulation may be burned, or portions of structures may be exploded. To avoid the possibility of lightning damage, the stipulations of the National Electrical Code should be observed. In particular, masts and metal structures that support antennas should be grounded by means of a directly–routed and unspliced ground wire

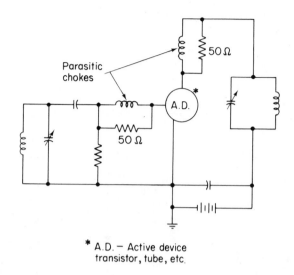

* A.D. – Active device
transistor, tube, etc.

Figure 9–8 Small chokes and/or resistors are used to suppress parasitics.

of ample size. A metallic underground water–piping system is recommended as a reliable ground connection. Each conductor of a lead–in
should be provided with a suitable lightning arrestor, unless the lead–in
is protected by a continuous metallic shield which is permanently
grounded. Exceptions can also be made in arrangements where the antenna itself is permanently and effectively grounded.

9–5 CB EQUIPMENT TROUBLESHOOTING—
GENERAL CONSIDERATIONS

Citizens band (CB) transmitters and receivers are basically similar to
amateur radio units. However, CB transmitters are comparatively low-
powered, ranging from 400 milliwatts to 5 watts. Figure 9–9 shows the
configuration for a 5–watt CB transmitter. The power rating is defined
as the product of DC voltage and current to the collector of the RF
power amplifier. Elaborate CB receivers are essentially the same as amateur superheterodyne receivers. Simple CB receivers often employ superregenerative detectors, as exemplified in Fig. 9–10. Portable CB equipment
is compactly constructed and hand–held, usually termed walkie–talkie
radios. One design distinction between CB and amateur receivers is that a
CB receiver may be fixed–tuned by means of quartz crystals.

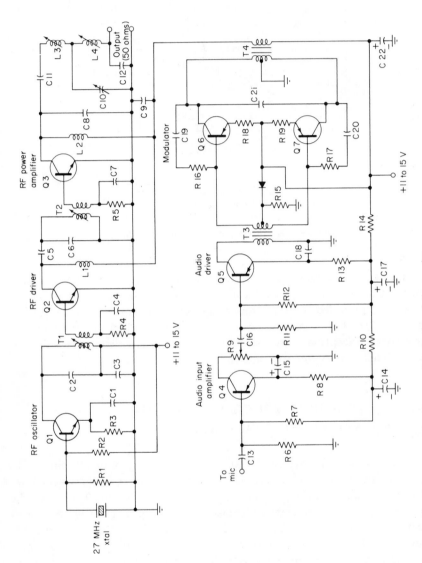

Figure 9–9 Configuration for a 27–MHz, 5-watt citizens band transmitter.

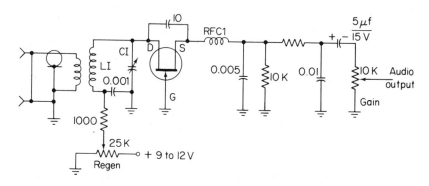

Figure 9–10 Basic superregenerative detector arrangement.

9–6 CB EQUIPMENT TROUBLESHOOTING PROCEDURES

Typical troubleshooting procedures for CB equipment are as follows:

Power Output. If the plate–power input to a CB transmitter is 5 watts, the power output is typically 3 watts. To check the power output, an RF wattmeter is utilized in place of the transmitting antenna. The wattmeter may or may not provide a 50–ohm load (dummy antenna). In any case, the power reading must be made across a 50–ohm resistive load. When the RF power output is subnormal, a systematic checkout of the transmitter is required. The power–type transistors are likely suspects, followed by leaky or open capacitors. DC–voltage measurements usually suffice to close in on a defective component.

Crystal Oscillator. In some cases, subnormal RF power output is caused by a defective quartz crystal. Off–frequency operation is likely to be an accompanying trouble symptom. Occasionally, a defective crystal will operate satisfactorily when the power is first applied, and then jump to another frequency after a while. Note that a replacement crystal must be designed for the circuit in which it is to be used. If a crystal is suspected of being defective, and the capacitors in the oscillator circuit are all right, it is advisable to check the adjustment(s) in the circuit before replacing the crystal. As an illustration, the adjustment of T1 in Fig. 9–9 would be checked. Frequency can be measured accurately with a CB frequency meter.

Modulator. If there is weak or no modulation of the RF carrier, or in the case of ample modulation but objectionably distorted output, there is a defect in the modulator section. However, before checking the modulator circuit, it is advisable to try another microphone of the same type. With reference to Fig. 9–9, the modulator stages are Q4, Q5, Q6 and Q7. An oscilloscope is the preferred test instrument, because it shows where distortion is occurring. Electrolytic capacitors are the most usual troublemakers, followed in frequency by faults in power transistors. If Q6, for example, develops appreciable collector–junction leakage, the output from the modulator will become weak and distorted.

Superregenerative Receiver. A superregenerative receiver (Fig. 9–10) is extremely sensitive, although it has the disadvantages of comparatively poor selectivity and of loud random noise output when an incoming signal is not present. The most common trouble symptom is lack of superregenerative action and operation as a conventional detector. Since a superregenerative circuit is quite critical, a systematic checkout of the detector–circuit components is usually required. DC–voltage measurements are less informative than in other circuitry, and technicians often make substitution tests of suspected components. Field–effect transistors are less likely to cause trouble, but a substitution check should be made if other components are not defective.

9–7 MOBILE EQUIPMENT TROUBLESHOOTING

Mobile radio receivers used in patrol cars, fire trucks, taxicabs, and other services are basically similar to FM broadcast receivers. However, bandwidths are considerably less: most mobile receivers have a bandwidth of approximately 3 kHz. The fidelity of voice reproduction can be compared with that of an ordinary telephone line. Another distinction of mobile equipment is its use of narrow–band FM (NFM) modulation. Thus, the deviation employed by mobile transmitters is in the range from 5 to 15 kHz. This is somewhat greater than the deviation used in most amateur NFM transmitters, which are limited to 3 kHz deviation. However, it is much less than the deviations used in FM–broadcast and TV–sound transmitters. The chief advantage of NFM communication over AM communication is its comparative immunity to static disturbances.

Figure 9–11 shows a block diagram for a VHF FM mobile receiver with a squelch circuit that switches off the audio amplifier unless there is an incoming signal that exceeds the noise level. A squelch circuit eliminates the loud random noise that is otherwise present when there is no incoming signal to quiet the discriminator system. This is an exam-

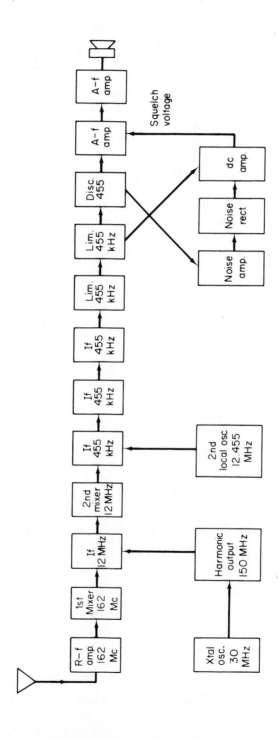

Figure 9-11 Block diagram for a **VHF FM** mobile receiver.

ple of a dual–conversion superheterodyne that employs crystal oscillators in the mixer stages. A typical squelch circuit that operates as an electronic switch is depicted in **Fig.** 9–12. It is triggered on or off by changes in the AVC voltage level.

Various types of frequency modulators are utilized in mobile transmitters, and it is instructive to consider the simple Varicap arrangement depicted in **Fig.** 9–13. The output frequency of an oscillator is varied in accordance with the instantaneous values of an audio waveform. This is accomplished by capacitance variation of diode junctions. As the voltage across the diodes increases, their junction capacitances decrease, and vice versa. Although there are differences in detail among the various designs of mobile radio transmitters and receivers, the basic troubleshooting approaches are the same as those explained previously for other classes of transmitters and receivers. All mobile equipment is provided with comprehensive instruction manuals, which should be referred to when puzzling trouble symptoms occur.

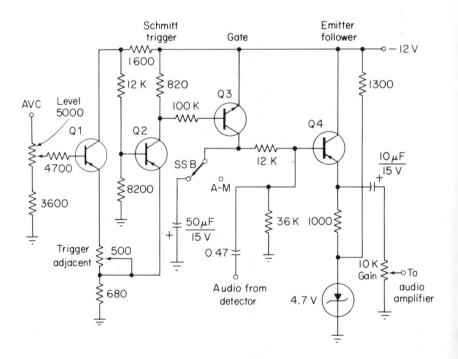

Figure 9–12 A squelch circuit that is actuated by AVC voltage.

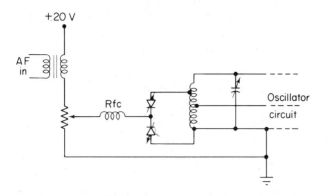

Figure 9–13 Basic Varicap frequency–modulation arrangement.

QUESTIONS

1. What type of troubleshooting procedures are involved in repairing radio transmitters?

2. What is one special requirement for servicing radio transmitters?

3. What are the causes of radio–receiver symptoms called birdies?

4. Why is alignment particularly critical in some high–frequency receivers?

5. What are several of the typical problems that are encountered in troubleshooting of amateur radio transmitters?

6. What is one main rule to follow when troubleshooting a transmitter that is disconnected from the supply voltage?

7. In Fig. 9–1, what is the purpose of the zener diode, CR1?

8. What is the purpose of diode CR2 in Fig. 9–1?

9. What is one of the problems encountered in measuring the DC voltages of an amateur radio receiver?

10. What is the first area of the receiver that should be checked if the receiver is "dead"?

11. In reference to Fig. 9–1, suppose a check of the DC voltage at R15 indicates zero volts. What is the problem?

12. What is the problem in Fig. 9–1 if a signal injected into point U1 gives a response, but a signal injected into point J1 gives no indication?

13. How can a suspected open capacitor be tested?

14. How can a suspected open coupling capacitor be tested?

15. What are the symptoms of an open decoupling or bypass capacitor?

16. What are some of the mechanical problems that can occur in a radio receiver? What are the causes of intermittent symptoms?

17. What is the advantage of the double–conversion receiver?

18. What is a simple method of determining whether front–end oscillations are present in a receiver?

19. How can you determine whether an oscillation is located in the IF section?

20. What are the probable symptoms if grid–leak resistor R4 in Fig. 9–3(a) is open?

21. What is the most probable cause of frequency drift during CW operation?

22. How can you determine that the cause of a frequency drift is located in the high–frequency oscillator?

23. What causes weak output from a radio receiver?

24. How are the IF amplifiers aligned in a radio receiver?

25. What are the symptoms of parasitic oscillations?

26. What is one of the most common causes of frequency drift?

27. How can parasitic oscillations be detected in a transmitter?

28. How can you avoid the problem of lightning damage to a transmitter?

29. How is the power rating of a CB transmitter defined?

30. How is the power output of a CB transmitter tested?

31. What are some of the most common troubles that cause reduced power output of a CB transmitter?

32. What are the characteristics of a superregenerative receiver?

33. What is the purpose of a squelch circuit in a mobile receiver?

10

ELECTRONIC ORGAN TROUBLESHOOTING

10-1 GENERAL CONSIDERATIONS

Although an electronic organ superficially appears different from a high–fidelity audio system, the similarities far exceed the technical distinctions. Its system organization is characterized by audio oscillators, frequency dividers, filter sections, and switching networks. This system has an appearance of considerable complexity, but this is due to the fact that comparatively simple circuits are employed dozens of times, often with progressive minor changes in component values. The external appearance of a modern electronic organ is shown in Fig. 10–1. Most organs utilize one or more external speakers (tone cabinets), although the smaller types of organs may have built–in speakers.

Electronic organ design varies somewhat, although most present–day designs contain five principal sections consisting of tone generators (audio oscillators), keying (switching) system, voicing (waveshaping) sections, audio amplifier, and speaker facilities. Figure 10–2 depicts these five basic sections. Note that the term repeat–percussion, or simply percussion, denotes a form of sound produced by plucking strings, striking cymbals, or beating drums. A block diagram for a typical electronic

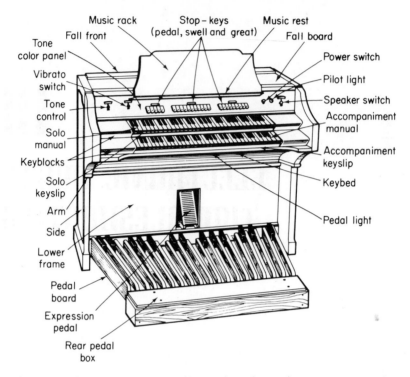

Figure 10–1 External appearance of a modern electronic organ.

organ is shown in Fig. 10–3. This is also a basic flow chart that indicates audio signal current paths.

Note that a tremulant section is essentially a modulator. That is, a tremolo consists of an amplitude modulation of an audio tone at a frequency of approximately 7 Hz. Again, a vibrato consists of a frequency modulation of an audio tone at a rate of approximately 7 Hz. A tibia is an organ tone (voice) that simulates flute tones. Traps are percussion instruments. FF denotes fortissimo (loudness of reproduction). Expression is an equivalent term for volume control. Most modern electronic organs employ modular design, with circuit boards that are easily replaceable, as exemplified in Fig. 10–4.

10–2 MAINTENANCE PROCEDURES

Elaborate electronic organs may operate with two or more tone cabinets, each of which contains bass, midrange, and tweeter speakers, as depicted

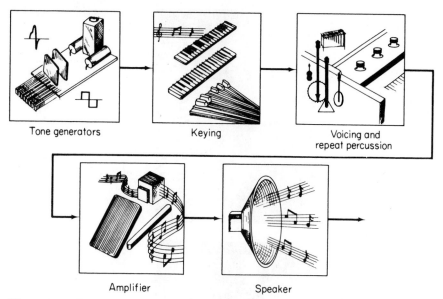

Tone generators   Keying   Voicing and repeat percussion

Amplifier   Speaker

Figure 10–2 Fundamental functions of a modern electronic organ.

in Fig. 10–5. In this example, potentiometers are provided to adjust the relative outputs from the speakers. These adjustments are made to suit individual preferences, and to compensate for the acoustic characteristics of the room in which the tone cabinets are placed. Note also that some tone cabinets contain a rotating vane which produces a vibrato–tremolo effect. The Baldwin–Leslie tone cabinet (Fig. 10–6) is representative; its bearings require occasional lubrication with a light machine oil. It is good practice to clean any accumulated dust and lint from the assembly before lubrication.

Manuals should be wiped occasionally with a soft damp cloth, with a slight amount of soap solution, if necessary. Stop keys (tabs) should also be wiped clean. Wood surfaces usually require dusting only, although scratches or dents must occasionally be contended with. Scratches are made less visible by application of furniture polish, or a scratch–stick may be utilized. Scratch–sticks are available in walnut, mahogany, and various lighter tones. Crayon-stain fillers can be used to fill deep scratches or dents. Technicians often carry cabinet–repair kits on assignments, so that professional touch–up procedures can be employed.

Various types of key switches and pedal switches are utilized in different types of electronic organs. Mechanical switches, exemplified in Fig. 10–7, are often used. Contacts require cleaning occasionally to avoid noisy action or no response. Contact wires require slight bending in some

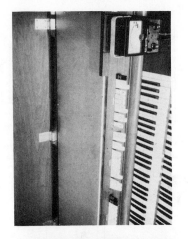

Figure 10–3 Block diagram for a typical electronic organ.

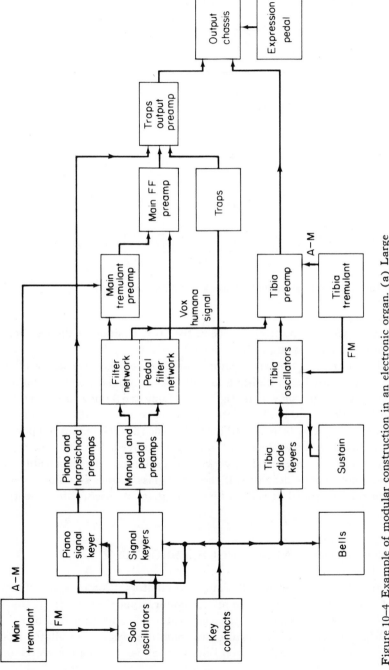

Figure 10–4 Example of modular construction in an electronic organ. (a) Large shield plate covers the circuit-board compartment; (b) Circuit boards are arranged in a rack; (c) Individual circuit boards slide out of slots in rack.

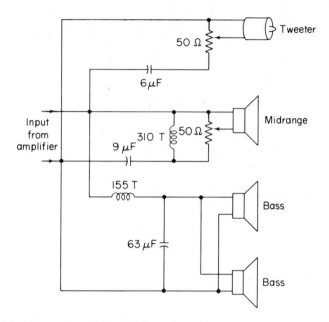

Figure 10–5 Example of individual speaker volume controls.

cases, to provide adequate clearance in the "up" position of the key, and adequate contact pressure in the "down" position.

Pilot lights and operating lamps must be replaced occasionally in maintenance procedures. Lighting facilities for a comparatively elaborate electronic organ are shown in Fig. 10–8. Tungsten lamps are generally used as pilot lights and to illuminate the pedal clavier (pedalboard). Small fluorescent tubes may be provided for the music rack. Note that the power circuit is fused. A slow–blow type is often employed. If a fuse blows, a short circuit has occurred; defective wire insulation may be responsible, or a lamp socket may be defective. If the inner glass surface of a fluorescent tube appears darkened, it should be replaced, as its useful life is nearly ended.

10–3 TONE–GENERATOR TROUBLESHOOTING

Tone generators are essentially audio oscillators. Complex–waveform output is usually provided, such as triangular, square, or distorted sine–wave shapes. A representative tone–generator configuration is shown in Fig. 10–9. A buffer stage is required to maintain accurate oscillator frequency under changing load conditions; that is, the load consists of

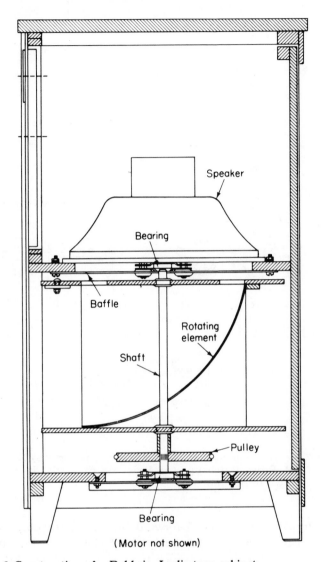

Figure 10–6 Construction of a Baldwin–Leslie tone cabinet.

various voicing filters (waveshaping circuits), which can be keyed in and out. An organ voice denotes a particular kind of tone, as explained in greater detail subsequently. Note that the oscillator depicted in Fig. 10–9 is not keyed, but operates continuously. However, diode CR1 is a keying diode for the vibrato function; that is, CR1 is ordinarily reverse-biased and is equivalent to an open circuit. On the other hand, when a

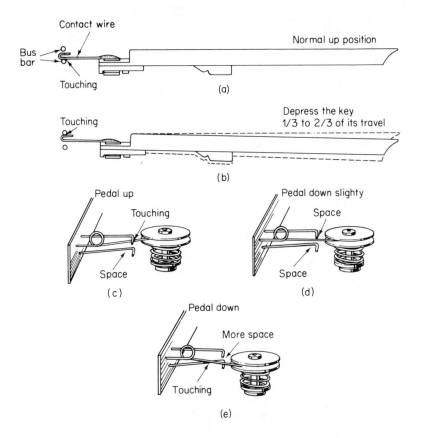

Figure 10–7 Keyswitch actions. (a) Key in resting position; (b) Key depressed; (c) Pedal switch up; (d) Pedal switch partly down; (e) Pedal switch fully down.

vibrato effect is desired, CR1 is forward–biased, and becomes equivalent to a low resistance, such as 100 ohms. This keying action permits a 7–Hz AC voltage from the vibrato section to enter the Q1A oscillator circuit, whereby a modulated tone is generated by the oscillator.

Troubleshooting a tone–generator circuit is generally approached by analysis of DC–voltage measurements. Observe that reference DC–voltage values can be determined by measuring voltages in another tone generator that is operating normally; that is, as tabulated in Fig. 10–9, a complement of tone generators employs progressively different L and C values in the same basic oscillator circuit. In turn, DC–voltage values are closely comparable. Malfunction is commonly caused by defective

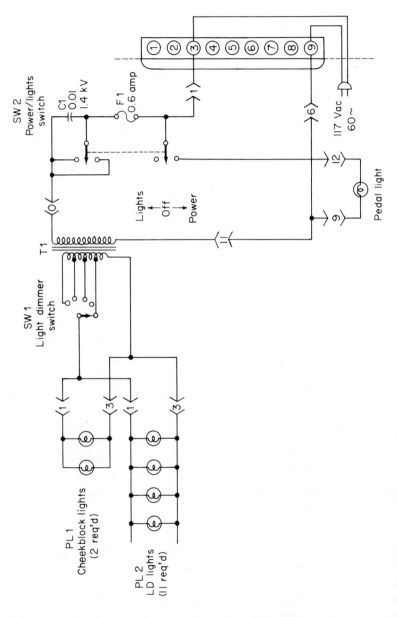

Figure 10–8 Lighting circuit for an electronic organ.

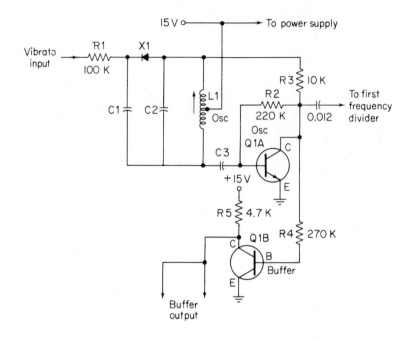

Figure 10–9 Representative tone–generator configuration (*Courtesy of* Rodgers Organ Co.).

Master oscillator tuning components				
Generators	C1	C2	C3	L1
C#–D	0.039	0.022	0.0068	P–9978–3
D#–E	0.033	0.018	0.0056	P–9978–3
F–F#	0.027	0.015	0.0047	P–9978–3
G–G#	0.022	0.012	0.0039	P–9978–2
A–A#	0.018	0.01	0.0033	P–9978–2
B–C	0.015	0.0082	0.0027	P–9978–2

capacitors. Transistors occasionally deteriorate, and diodes may develop a poor front–to–back ratio. Note that if CR1 has a poor front–to–back ratio, the vibrato tab switch will cipher. This means that the vibrato action continues although the vibrato tab is switched off. If a tone generator operates off–frequency, the trouble is usually due to a marginal component defect, such as a slightly leaky capacitor. No attempt should be made to tune an electronic organ until all component defects have been corrected. Observe that a slug is provided in L1 for tuning the oscillator.

10–4 TROUBLESHOOTING THE VOICING SECTION

A voicing section processes the waveforms from the oscillator section, and shapes the arbitrary input (such as a square wave) into an output that has the characteristics of a musical tone. Another aspect of the voicing process is frequency division, to produce an output waveform that has a lower fundamental frequency than the input waveform. Waveshaping is accomplished by means of RC or RLC filters, termed formant filters, as exemplified in Fig. 10–10. Frequency division involves triggered or synchronized oscillators. Thus, a 1046.6–Hz C note is high-pitched, but when processed by a 2–to–1 frequency divider, produces a 523.3–Hz C note which is an octave lower in pitch. Note that pitch corresponds to the fundamental frequency of a voicing waveform.

Sustain voicing denotes a tone that diminishes gradually in its intensity after a key has been released. This effect is accomplished by

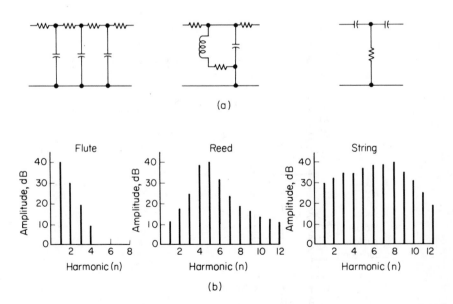

(a)

(b)

Figure 10–10 Typical formant filters, and output frequency spectra. (a) RC low–pass, RLC bandpass, and RC high–pass sections; (b) Frequency spectra for flute, reed, and string waveforms.

means of capacitor–charge decay, as exemplified in Fig. 10–11. The
diodes operate as switches, and are actuated by DC voltages. If a diode
becomes defective, certain operating symptoms are produced, as follows:
If X2 is "shorted" or leaky, the note will cipher when the long–sustain
function is turned off as explained below. On the other hand, if X2 is
"open," no sustain action is obtained, because C1 cannot charge. Or, if
X3 becomes leaky or "shorted," abnormal sustain action takes place.
Again, if X3 is "open," no sustain action is obtained, because C1 cannot
discharge through the tone generator.

Figure 10–12 shows how the voltage to the tone generator normally
decays slowly during sustain voicing. Long–sustain action is initiated at
a higher voltage than short–sustain action. The switching diodes in Fig.
10–12 correspond to X2 and X3 in Fig. 10–11. If a contact–multiplying
diode (Fig. 10–13) is "open," the associated keyer becomes inoperative.
On the other hand, if a contact–multiplying diode is "shorted," the stops
(tab switches) associated with that sustain function will also be turned
on whenever the stop associated with the "shorted" diode is turned on.
Observe in Fig. 10–13 that 4-foot and 2-foot piccolo voices are indicated.
This is a terminology that has been carried over from pipe–organ tech-

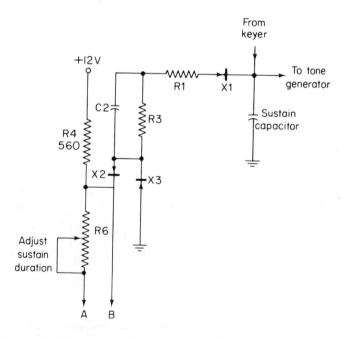

Figure 10–11 Typical sustain–voicing configuration.

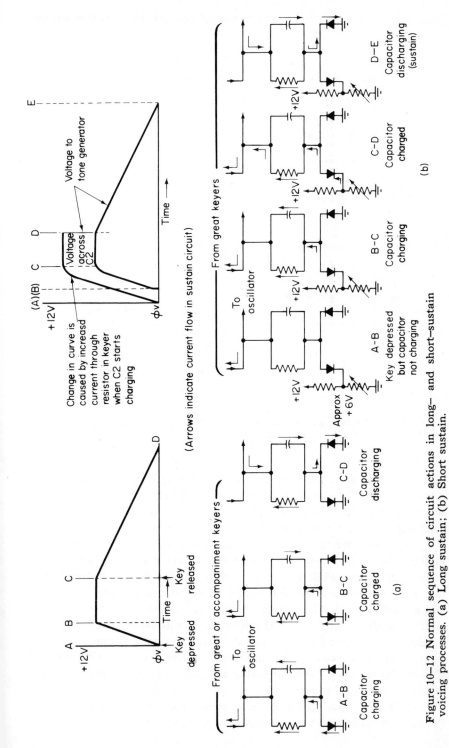

Figure 10–12 Normal sequence of circuit actions in long– and short–sustain voicing processes. (a) Long sustain; (b) Short sustain.

225

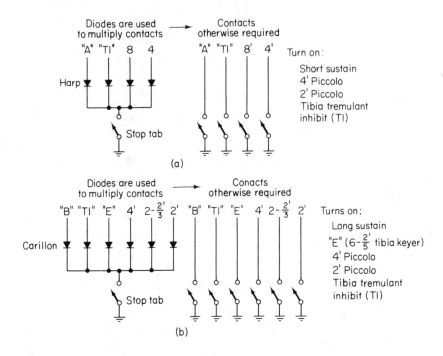

Figure 10–13 Contact–multiplying diodes employed in sustain voicing. (a) Harp stop tab; (b) Carillon stop tab.

nology, in which the pitch of a tone was denoted by the length of the organ pipe that produces that pitch. As an illustration, a 16–foot pitch has a fundamental frequency of 32.5 Hz, and an 8–foot pitch has a fundamental frequency of 65 Hz.

Next, consider the tremulant/vibrato voicing generator exemplified in Fig. 10–14. This generator produces a 7–Hz sine–wave output, and runs continuously while the organ is in operation. Note that the tremulant stop tab in this example turns on a light–dependent resistor (LDR) which switches the AM output. Again, the vibrato stop tab turns on another LDR to switch the FM output. An oscilloscope is very useful to locate a faulty circuit branch, and defective components can usually be pinpointed by DC voltage and/or resistance measurements. Observe that if diode X1 is "open," the harpsichord stop switch cannot be turned on, but if the diode is "shorted," the switch cannot be turned off. Similarly, if diode X2 is "open," the piano stop switch cannot be turned on, but if the diode is "shorted," the switch cannot be turned off.

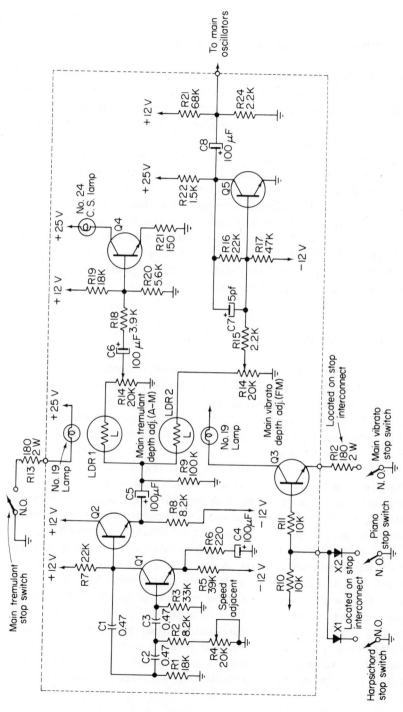

Figure 10–14 Configuration of a typical tremulant/vibrato voicing generator.

10-5 AMPLIFIER TROUBLESHOOTING

Troubleshooting procedures for electronic–organ amplifiers are basically the same as explained in Chapter 4 for hi–fi stereo amplifiers. Note, however, that an organ amplifier may or may not be a hi–fi amplifier. That is, the basic function of an organ is to *produce* musical tones, whereas the basic function of a hi–fi system is to *reproduce* musical tones. Elaborate organs contain an "entertainment center" which provides reproduction of tape recordings and FM programs. In turn, hi–fi amplifiers are utilized in this section of the organ. A typical electronic–organ amplifier configuration is shown in Fig. 10–15. Its frequency range is less than in hi–fi systems, and is determined in part by the feedback loop through C6 and R8. Thus, if C6 becomes leaky or "open," the tones that are produced become distorted. Or, if C6 becomes "shorted," distortion results with abnormal gain. Figure 10–16 shows the general relationship between percentage negative feedback and amplifier gain.

10-6 ELECTRONIC ORGAN TUNING

Tuning is sometimes done by ear, and a technician with a good musical background can tune an organ properly by ear. However, non–musical technicians need suitable instruments or devices to tune an organ accurately. Table 10–1 lists the frequencies of the tempered scale, which is used in the great majority of organs. A flute voice is preferred during the tuning procedure, because it has an approximate sine waveform, and its pitch is easier to evaluate than that of other organ voices. The standard starting point is the 523.25–Hz C tone, which is usually established by a tuning fork. Only one octave need be tuned in many organs, because the remaining octaves are synchronized by frequency dividers. However, some organs may employ more than 150 individual tone generators.

　　　　It is instructive to consider the tuning procedure for a typical organ that utilizes frequency dividers. A so–called rough–tuning procedure is followed first, and terminated with a fine–tuning procedure. The C note is first established by means of a tuning fork, or by the C note from a companion instrument. Then the beat principle is employed to tune the remaining tone generators by ear; that is, when two notes are sounded simultaneously, the number of beats is counted over a period of 10 or 30 seconds. For example, if C3 and G3 are sounded simultaneously, there will be 9 beats produced in 10 seconds, or 27 beats in 30 seconds, pro-

228

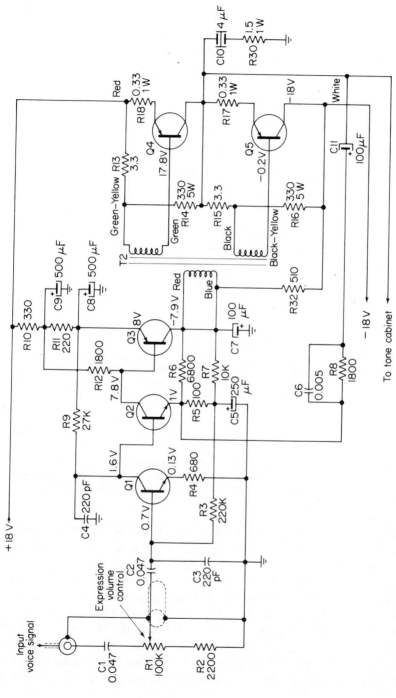

Figure 10–15 Typical electronic–organ amplifier configuration.

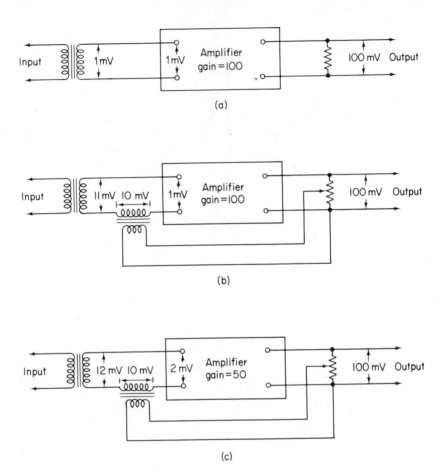

Figure 10–16 Relation of percentage negative feedback to amplifier gain. (a)
No negative feedback; (b) Ten percent negative feedback; (c)
Ten percent feedback, at 50% maximum available gain.

vided the pitches are correct. In the example of Fig. 10–17, the pitch of
each tone generator is adjusted by means of a slug in the associated
oscillator coil.

Table 10–2 shows the rough-tuning procedures followed in this example, and Table 10–3 lists the fine-tuning procedures. If a "flag" is
taped on the aligning tool, as shown in Fig. 10–2, it is easier to observe
the amount that a slug has been turned. Using the flute voice, and the
C3 note zero-beat against a tuning fork, the B2 note is next zero beat
against the C3 note, and rough-tuned by turning the B2 alignment tool
1¾ turns clockwise. This basic rough–tuning procedure is started

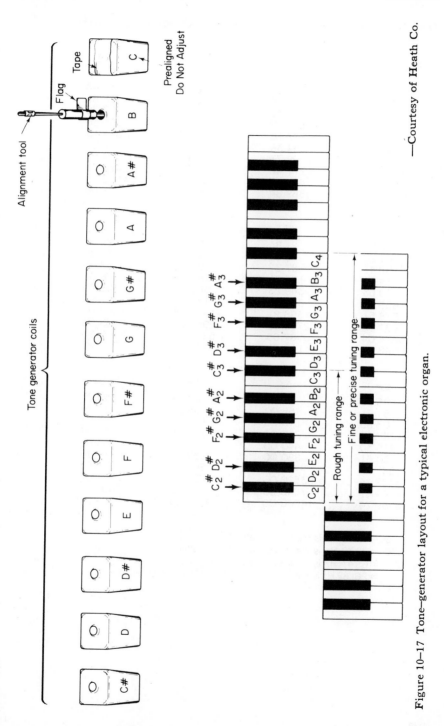

Figure 10–17 Tone–generator layout for a typical electronic organ.

—Courtesy of Heath Co.

Table 10–1 Frequencies of the Tempered Scale (Frequencies in Hertz).

C	C#	D	D#
16.35			
32.70	17.32	18.35	19.44
65.41	34.65	36.71	38.89
130.81	69.30	73.42	77.78
261.62	138.59	146.83	155.56
523.25	277.18	293.66	311.13
1046.50	554.36	587.33	622.25
2093.00	1108.73	1174.66	1244.51
4186.01	2217.46	2349.32	2489.01
8372.02	4434.92	4698.64	4978.03
16744.03	8869.84	9397.27	9956.06

E	F	F#	G
20.60	21.83	23.12	24.50
41.20	43.65	46.25	49.00
82.41	87.31	92.50	98.00
164.81	174.61	185.00	196.00
329.63	349.23	369.99	392.00
659.26	698.46	739.99	783.99
1318.51	1396.91	1479.98	1567.98
2637.02	2793.82	2959.95	3135.96
5274.04	5587.65	5919.90	6270.93
10548.08	11157.30	11839.81	12541.86

G#	A	A#	B
25.96	27.50	29.14	30.87
51.91	55.00	58.27	61.74
103.83	110.00	116.54	123.47
207.65	220.00	233.08	246.94
415.30	440.00	466.16	493.88
830.61	880.00	932.33	987.77
1661.22	1760.00	1864.65	1975.53
3322.44	3520.00	3729.31	3951.06
6644.87	7040.00	7458.62	7902.13
13289.74	14080.00	14917.23	15804.26

CCCC= 16.35 Hz is the lowest note of 32 ft pitch
CCC = 32.70 Hz is the lowest note of 16 ft pitch
CC = 65.41 Hz is the lowest note of 8 ft pitch
C = 261.62 Hz is the popularly termed middle C of the keyboard

Table 10–2 Rough-Tuning Procedure for the Example of Figure 10–17 (*Courtesy of* Heath Co.).

Zero beat	Then adjust generator coil clockwise	
1. (✓) B2 with C3	B	1–3/4 turns
2. (✓) A#2 with B2	A#	2–1/4 turns
3. ( ) A2 with A#2	A	1–3/4 turns
4. ( ) G#2 with A2	G#	2 turns
5. ( ) G2 with G#2	G	2–1/4 turns
6. ( ) F#2 with G2	F#	1–3/4 turns
7. ( ) F2 with F#2	F	2 turns
8. ( ) E2 with F2	E	2–1/4 turns
9. ( ) D#2 with E2	D#	1–3/4 turns
10. ( ) D2 with D#2	D	2 turns
11. ( ) C#2 with D2	C#	2–1/4 turns

Table 10–3 Fine-Tuning Procedure for the Example of Figure 10–17 (*Courtesy of* Heath Co.).

Play notes	Turn clockwise	Beats in 10 seconds	Beats in 30 seconds
1. ( ) C3 and G3	G	9	26
2. ( ) G2 and D3	D	7	20
3. ( ) D3 and A3	A	10	29
4. ( ) A2 and E3	E	7	23
5. ( ) E3 and B3	B	11	33
6. ( ) B2 and F#3	F#	8	25
7. ( ) F#2 and C#3	C#	6	18
8. ( ) C#3 and G#3	G#	10	28
9. ( ) G#2 and D#3	D#	7	21
10. ( ) D#3 and A#3	A#	11	31
11. ( ) A#2 and F3	F	8	24
12. ( ) F2 and C3	Check only.	6*	18*

NOTE: Step #12 is a check of how accurately the tuning was performed. If the indicated number of beats are heard in this check, the Organ is in perfect tune. If the beats counted are within ± 3 beats of the indicated beats, it is acceptable. The number of beats that you hear above or below the indicated beats represents the total tuning error accumulated during fine adjustment of the eleven coils.

* ±3 beats are acceptable on this check. If more or less, repeat all the above steps.

using the trombone voice. As listed in Table 10–3, beats may be counted over 10–second or 30–second intervals. Greater accuracy usually results from the use of 30–second intervals. A stop watch may be used, or a conventional watch with a sweep–second hand. Note that the alignment tool should always be removed from a coil before starting to count beats in a fine–tuning procedure.

Various electrical and electronic organ–tuning aids are available, and are often utilized by technicians. For example, the Conn Strobotuner shown in Fig. 10–18 has been used for years. It employes a stroboscopic disk, which is illuminated by an intermittent light source synchronized by the acoustic output from the organ. A microphone is used as a transducer, followed by an amplifier. The stroboscopic disk is driven by a synchronous motor, and when the organ pitch is correct, a specified octave band on the disk will appear to stand still. However, if the pitch is too high or too low, the band will appear to turn clockwise or counterclockwise. Considerable time is saved in practice, since the results of a tuning adjustment is evident at once, and no delay is involved in counting beats.

Figure 10–18 Appearance of the Conn Strobotuner.

QUESTIONS

1. What is a tremulant section?
2. What does a vibrato consist of and what is the audio tone rate?
3. What are traps used for in an electronic organ?
4. What is a tibia?
5. What is expression an equivalent term for?
6. What should you do before lubricating a bearing assembly?
7. Why should contacts be cleaned occasionally?
8. How can you tell that a fluorescent tube is defective?
9. What is the purpose of a buffer stage?
10. What is the best method for determining the exact DC voltage values of a tone generator?
11. What is the purpose of the voicing section?
12. What symptoms will you observe if the diode CR in Fig. 10–11 is leaky or shorted?
13. Where does the term piccolo voices originate?
14. What are the symptoms if the diode CR in Fig. 10–14 is open?
15. Where is the standard starting point on the scale for tuning an organ, and how is this point established?
16. How is the pitch of each tone generator, in Fig. 10–17 established?

11

DIGITAL COMPUTER TROUBLESHOOTING

11-1 GENERAL CONSIDERATIONS

Electronic digital computers use various switching arrangements to perform mathematical operations such as addition, subtraction, multiplication, and division. Transistors and diodes are commonly employed as electronic switches. Many millions of elementary operations can be performed per second. Figure 11–1 shows the appearance of a typical digital computer. Modular construction, as exemplified in Fig. 11–2, is standard practice. This design minimizes down–time, because a fault need only be localized in a module, which is quickly replaced. Then the defective component can be pinpointed and the module repaired at any convenient time. Although the configuration of a digital computer appears to be enormously complex, comparatively simple circuits are used to perform electronic switching functions, and these "building blocks" are arranged in series for sequential operation. This is a fortunate situation for the computer troubleshooter, because it enables him to analyze digital circuitry with comparatively simple test equipment.

One of the basic "building blocks" is the bistable multivibrator, or flip–flop, depicted in Fig. 11–3. If Q1 is conducting, Q2 is nonconducting, and vice versa. In turn, two input pulses are required to complete one

Figure 11–1 Appearance of a modern digital computer (*Courtesy of* IBM Corp.).

cycle of operation. P1 and P2 denote the trigger–pulse relation to the output waveform. Since the transistors are biased beyond cutoff, a flip–flop will remain in one state indefinitely until such time that a trigger pulse is applied. Because a square wave is generated for each pair of incoming trigger pulses, a flip–flop is said to divide by two; that is, a flip–flop produces half as many output pulses as the number of applied trigger pulses. Note that the 220–PF capacitors in Fig. 11–3 are speed–up capacitors. They are so called because they provide fast trigger action by passing the higher harmonics of a pulse over the coupling resistors. Thereby, the circuit response is not slowed down by integrating action of the junction capacitances. Note that if a speed–up capacitor becomes open–circuited, the response of the flip–flop will be slowed down.

Flip–flops are connected in a chain for binary storage of digital pulses. The chain is called a counter. Since digital computers operate on the basis of the two states—*on* and *off* (0 and 1)—our conventional decimal system of numbers that employs ten basic states cannot be utilized. Therefore, a computer uses the binary system of numbers, which is based on powers of 2 instead of powers of 10. As an illustration,

Figure 11–2 Solid–state module used in a digital computer (*Courtesy of* Computer Control Co., Inc.).

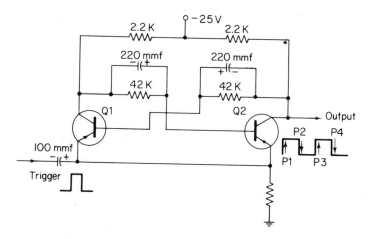

Figure 11–3 Basic flip–flop circuit.

239

in the decimal system, the number 397 can be expressed in the form $(3 \times 10^2) + (9 \times 10^1) + (7 \times 10^0)$. Observe that the position of a digit corresponds to a power of 10, and the value of the digit indicates how many times this power of 10 is to be added to itself. For instance, in the number 30, the position of 3 corresponds to the first power of 10, and the value of 3 indicates that 10 (10^2) is to be added to itself three times.

In the binary system, all numbers are expressed in terms of the digits 0 and 1. With reference to a flip–flop, 0 and 1 correspond to its *off* and *on* states. As an illustration, *on–off–on*, or 101 in binary notation, can be expressed as $(1 \times 2^2) + (0 \times 2^1) + (1 \times 2^0)$ which is equal to 5 in decimal notation. Figure 11–4 shows how the decimal numbers up to 5 are counted in binary form by a flip–flop chain. The input triggers are stored in the chain progressively as they are counted. Indicating lamps may be used for readout of the binary numbers. Note that the signal flow is depicted as progressing from right to left in Fig. 11–4, so that the binary numbers will be indicated in conventional order. After five input triggers have been applied, the state of the counter corresponds to the binary number 101. If it is desired to reset the counter back to 000 at any time, a pulse is applied to the first transistor in each flip–flop, thereby driving it into conduction.

Another digital–computer "building block" is the logic switch. These switch arrangements are used in adders, which are fundamental sections of calculating circuits. (Addition is the basic function of a digital computer.) Binary addition is an operation that follows from the definition of binary digits. For instance, if we wish to add 5 to 3 with a sum of 8, the binary operation will add 0101 and 0011 with a sum of 1000. The truth table (addition table) for two binary numbers is written:

$$0 + 0 = 0$$
$$0 + 1 = 1$$
$$1 + 0 = 1$$
$$1 + 1 = 0, \text{ and 1 to carry}$$

A simple binary adder with its logic–switch elements is depicted in Fig. 11–5. The truth table for the adder is the same as explained above. Note that the AND blocks have no output unless two 1 inputs are present. However, the OR block will have an output if a 1 pulse is present at either input. The NOT block has a 1 output if there is a 0 input, and it has a 0 output if there is a 1 input. Figure 11–6 exemplifies typical logic switches, also called gates. Note that the diodes are forward–biased in

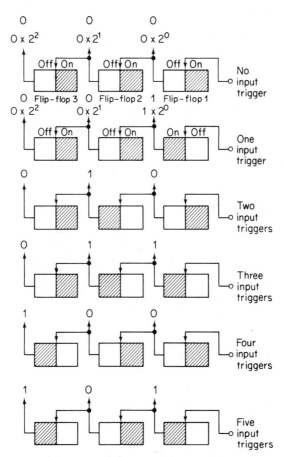

Figure 11–4 Development of binary numbers by a flip–flop chain.

both the OR and the AND gates. The input pulse amplitude is sufficient to cut off a diode. A NOT gate is typically a transistor operated in the CE mode, which inverts the polarity of an input pulse in its output circuit. Conventional gate symbols used in computer calculating circuits are shown in Fig. 11–7.

The basic binary adder depicted in Fig. 11–5 is generally called a half adder. It can be compared with a flip–flop, in that half adders can be connected in a chain as in Fig. 11–4. However, a half–adder chain is much faster than a flip–flop chain in processing binary numbers. That is, a half–adder chain is energized by two inputs at the same time, whereas a flip–flop chain must be energized by one input at a time. Thus, the half–adder chain is twice as fast. If addition is to be performed by a

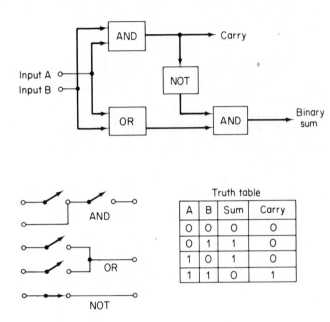

Figure 11–5 A basic binary adder with its logic–switch circuits.

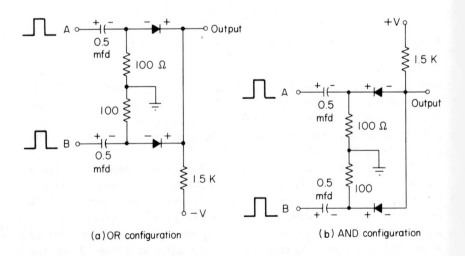

(a) OR configuration   (b) AND configuration

Figure 11–6 Examples of diode logic switches, or gates.

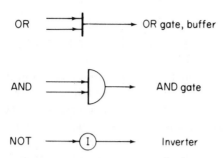

Figure 11–7 Conventional computer gate symbols.

flip–flop chain, the first binary number is fed into the chain, followed by the second binary number. If the inputs must accommodate a carry from the previous position, a full–adder configuration is utilized, as depicted in Fig. 11–8. Nearly all present-day calculating circuitry employs integrated circuits which perform the same basic functions as circuits with discrete components.

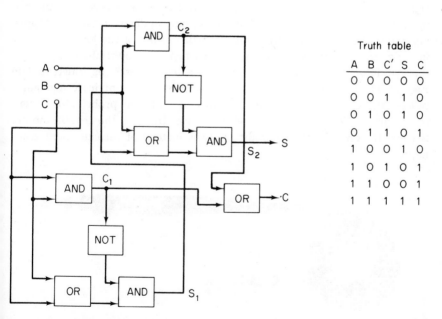

Figure 11–8 Full–adder arrangement consists of two half adders and an **OR** logic switch.

11-2 TROUBLESHOOTING DIGITAL COMPUTERS

A logic probe, illustrated in Fig. 11-9, is the most important single test instrument used in digital–computer troubleshooting. It is employed to trace logic levels and pulses through integrated circuitry to determine whether the point under test is logic high, low, bad level, open–circuited, or pulsing. The probe has preset logic thresholds of 2.0 and 0.8 volts, which correspond to the high and low states of transistor–transistor-logic (TTL) and diode–transistor–logic (DTL) circuits. When touched to a high–level point, a bright band of light appears around the probe tip. When touched to a low–level point, the light goes out. Open circuits or voltages in the "bad level" region between the preset thresholds produce illumination at half brilliance. Single pulses of 10 ns or greater widths are made readily visible by stretching to one–twentieth second. The lamp flashes on or blinks off depending upon the polarity of the pulse. Pulse trains up to 50–MHz repetition rate cause the lamp to blink off and on at a 10–Hz rate.

Figure 11-10 shows the logic probe in use. The circuit under test can first be run at normal speed, while checking for the presence of key signals, such as clock, reset, start, shift and transfer pulses, as explained in digital–computer courses. Next, the circuit under test can be stepped one pulse at a time while checking the truth tables of the logic packages, in order to turn up any defects. In this type of troubleshooting procedure, it is very helpful to also employ a logic pulser, as illustrated in Fig. 11-11. The pulser provides a convenient means of injecting single pulses, the effects of which are monitored with the probe. No adjustments are made on the probes while in use. In turn, the troubleshooter is freed to concentrate on circuit analysis instead of measurement techniques.

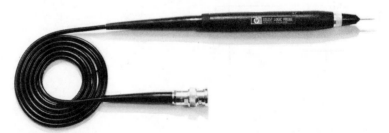

Figure 11-9 Logic probe used to troubleshoot digital circuitry (*Courtesy of* Hewlett–Packard Co.).

Figure 11–10 Logic probe in use (*Courtesy of* Hewlett–Packard Co.).

Connectors provided with the logic probe facilitate connection to the required 5–volt supply from either the circuit under test or a laboratory supply. A ground clip is provided which may be used to directly connect the probe circuitry to the ground of the circuit under test. Figure 11–12 shows a block diagram of the logic–probe circuitry and its response to different inputs.

When using the logic pulser shown in Fig. 11–11, its tip is touched to the circuit under test, and the pulse button is pressed. In turn, all circuits connected to the node (outputs as well as inputs)·are briefly driven to their opposite state. No unsoldering of IC outputs is required. Since the pulse injection is automatic, the troubleshooter need not con-

Figure 11–11 Logic pulser used to troubleshoot digital circuitry (*Courtesy of* Hewlett–Packard Co.).

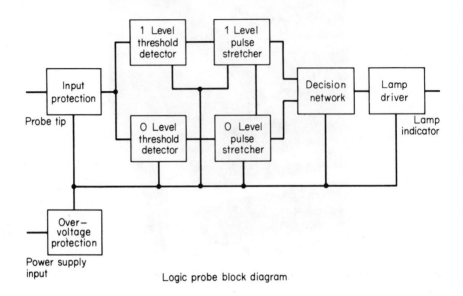

Logic probe block diagram

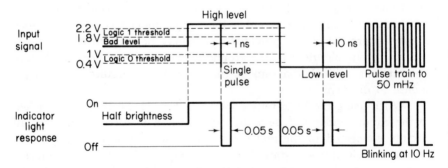

Probe response to different inputs

Figure 11–12 Block diagram of the logic–probe circuitry and its response to different inputs (*Courtesy of* Hewlett–Packard Co.).

cern himself with whether the test node is in the high or low state. High nodes are pulsed low, and low nodes high, each time that the button is pressed. The pulser is essentially a single–shot pulse generator in probe form. An ability to source or sink up to 0.65 ampere ensures sufficient current to override IC outputs in either the high or low state. An output pulse width of 0.3μs limits the amount of energy delivered to the device under test, thereby eliminating the possibility of destruction.

Another very useful test instrument is the logic clip, illustrated in Fig. 11–13. This unit clips onto TTL or DTL ICs and instantly displays the logic states of all 14 or 16 pins. Each of the clip's 16 light–emitting diodes independently follows level changes at its associated pin; a

(a)

(b)

Figure 11–13 Logic clip used to troubleshoot digital circuitry. (a) Appearance of clip; (b) Logic clip in use (*Courtesy of* Hewlett–Packard Co.).

lighted diode corresponds to a high logic state. The clip contains its own gating logic for locating the ground and the $+5$ volt V_{cc} pins. A logic probe is much easier to use than an oscilloscope or a voltmeter when the question is whether a lead is in the high or low (1 or 0) state. Pulser–probe and pulser–clip combinations enable the troubleshooter to quickly answer such questions as "Is a gate functioning?", "Is a pin shorted to ground or V_{cc}?", and "Is a counter counting?" without unsoldering pins or cutting printed–circuit conductors.

Gate operation is tested with the pulser driving the input and the probe monitoring output pulses. When pulses do not appear at the output, the pulser and probe can be applied to the same pin to determine if the failure is due to a short to ground or V_{cc}. Testing sequential circuits such as flip–flops is accomplished with the clip simultaneously monitoring all output states while the pulser applies reset pulses to the device. Faulty operation becomes immediately apparent since the IC will not go through its prescribed sequence of states.

An oscilloscope is also essential in digital–computer troubleshooting. For example, erratic operation can be caused by a seriously distorted pulse waveform, as exemplified in Fig. 11–14. Similarly, pulse repetition rates and pulse amplitudes can be accurately measured with a calibrated scope. Triggered sweep is an essential feature, and the vertical amplifier must have good 10–ns pulse response. High sensitivity is not required, because pulse amplitudes are normally in the order of 2 volts peak–to–peak. A low–capacitance probe is necessary to provide high input impedance to the scope, and thereby avoid objectionable circuit loading. DC response is desirable, although the majority of tests in calculating circuits can be made with an AC oscilloscope.

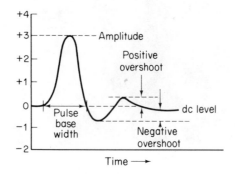

Figure 11–14 A seriously distorted digital pulse.

QUESTIONS

1. What components are usually employed as electronic switches?
2. What are the characteristics of a flip–flop circuit?
3. What type of waveform does a flip–flop circuit generate?
4. How are flip–flop circuits connected to produce a counter?
5. What is the most important single piece of test equipment used in the troubleshooting of a digital computer?
6. What is the purpose of the pulser?
7. What is a pulser?
8. How is burnout of circuits prevented when a pulser is used?
9. How is the logic state of an IC displayed with the logic clip?
10. What is the purpose of a low–capacitance probe?

12

ELECTRONIC INSTRUMENT TROUBLESHOOTING

12-1 GENERAL CONSIDERATIONS

Electronic instruments require occasional checkups to ensure that they are operating within specifications. Sometimes a catastrophic component failure occurs which produces a "dead–instrument" symptom. Marginal component failures cause abnormal operation or inaccurate indication which may or may not be apparent in the absence of a routine checkup. Indicating instruments such as VOMs and TVMs should be checked at intervals for calibration accuracy. Oscilloscopes should be checked for time–base and vertical–deflection voltage calibration. It is often advisable also to check oscilloscopes for deflection linearity. Signal generators need to be checked for calibration accuracy and output voltage level. In addition, a sweep–frequency generator should be checked for uniformity of output over the swept band. Square–wave and pulse generators require checking for rise time and accuracy of repetition–rate calibration. Color-bar generators should be checked for subcarrier calibration accuracy, phase accuracy of the various outputs, and output voltage level.

12-2 VOM MAINTENANCE AND CALIBRATION

A VOM (Fig. 12–1) is designed with a balanced pointer, so that the zero indication will normally be correct whether the instrument is operated in a vertical or a horizontal position. However, if the pointer becomes slightly unbalanced, the pointer may shift more or less off–zero when the instrument is moved from a vertical to a horizontal position, or vice versa. In such a case, the zero–set screw below the scale plate should be adjusted slightly, as required. As exemplified in Fig. 12–2, the ohmmeter section of a VOM employs batteries. These batteries must be replaced when the ohmmeter is no longer within calibration. Note that ohmmeter calibration can be checked by means of one or more precision resistors with a tolerance of ±1%; that is, the accuracy rating of service–type ohmmeters is in the order of ±3% (based on the arc of indication error as shown in Fig. 12–3).

DC–voltage calibration of a VOM is checked to best advantage with mercury cells or batteries. A mercury battery is marked for rated voltage, and this value is comparatively accurate and stable over the useful life of the battery. The tolerance on mercury–battery voltage ratings is in the same order of magnitude as the tolerance on service–type VOM

Figure 12–1 Simpson volt–milliammeter Model 261 (*Courtesy of* Simpson Electric Co.).

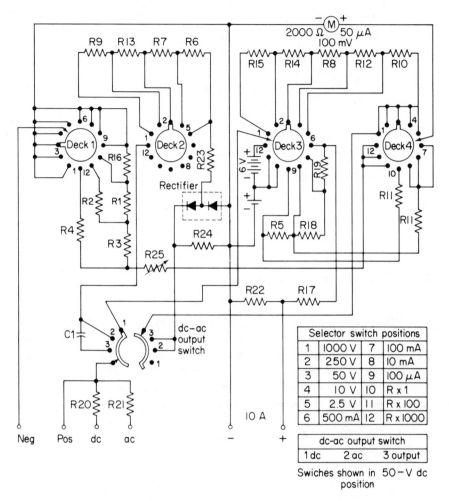

Figure 12–2 Simpson Model 260 VOM configuration (*Courtesy of* Simpson Electric Co.).

accuracy. In general, it is impractical to check the **DC–voltage** accuracy of a VOM with a standard cell such as Weston cell, because the current demand of the instrument is excessive. Note that the **AC–voltage** calibration of a VOM can also be checked to good advantage with mercury cells or batteries. Observe carefully that the AC voltage scale will read 2.22 times the DC source voltage if a half–wave instrument rectifier is used, and the AC scale will read 1.11 times the DC source voltage if a full–wave instrument rectifier is utilized in the VOM. In the example of Fig. 12–2, half–wave rectification is employed.

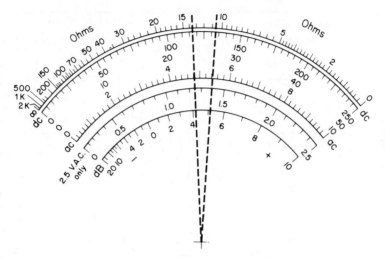

Figure 12–3 An example of arc of indication error.

VOM calibration is often impaired as a result of accidental over-loads, which may change the values of multiplier resistors, and/or bend the pointer in the mechanism. Excessive vibration or shock can damage the jeweled bearings and weaken the field of the permanent magnet. Overload on the AC–voltage function can damage the instrument recti-fiers. If a VOM is stored with weak or dead batteries, leaking chemicals may damage switches and other metallic components. When a VOM is found to be out of tolerance, it is usually advisable to send the instru-ment to a specialized repair depot. Special facilities and tools are re-quired to repair meter mechanisms and to remagnetize permanent magnets.

12–3 TVM MAINTENANCE AND CALIBRATION

A transistor voltmeter (TVM) or transistor volt–ohmmeter (TVOM) (Fig. 12–4) has an appearance somewhat similar to a VOM. However, the sensitivity of a TVM is much greater as a result of current amplifica-tion by a transistor bridge, as exemplified in Fig. 12–5. Note that diodes D3 and D4 are protective devices which ensure that the meter mecha-nism will not burn out if the instrument is accidentally overloaded. The more elaborate types of VOMs also employ protective diodes. Diode CR5 is a temperature–compensating diode, which maintains the gain of Q1 and Q2 constant over a considerable range of temperature. Note that

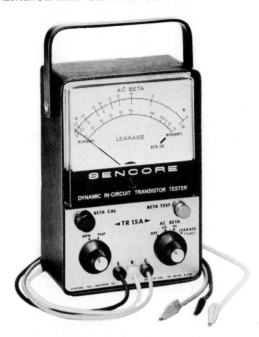

Figure 12–4   Sencore Model FE14 FET meter (*Courtesy of* Sencore).

input resistance of the TVM in this example is 15 megohms on all ranges. Therefore, the DC calibration can be checked directly with a standard cell such as a Weston cell, inasmuch as the current demand is less than a microampere.

Since a TVM or TVOM is battery–operated, it is necessary to check the batteries at the outset, in case the indication accuracy is out of tolerance. Transistors are normally very long lived, but may eventually require replacement. In such a case, it may be necessary to readjust the calibration control. For example, note the 5k "DC cal" control between Q1 and M1 in Fig. 12–5. It may also be desirable or necessary to read-just the DC balance control, in case the zero–adjust control is out of range. Note in passing that although diodes D3 and D4 protect the meter movement against burnout, there is no protection against damage to multiplier resistors in the case of heavy overload. Therefore, the multiplier resistors should be checked at the outset in case the instrument goes out of calibration after accidental overload.

Figure 12–6 shows a typical configuration for the AC voltage–measuring function of a TVOM. All instruments of this type provide both rms and peak–to–peak scales, which facilitate checking of AC calibration. For example, a standard cell such as a Weston cell or a

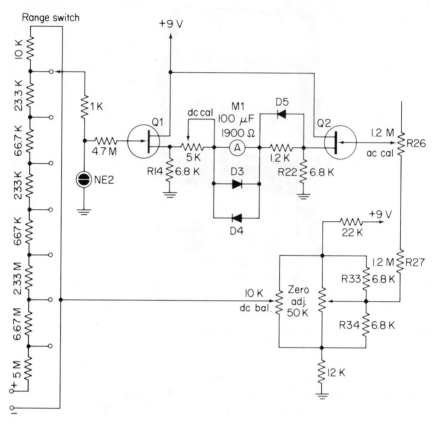

Figure 12–5 Basic **TVM** configuration, with field–effect transistor bridge.

mercury battery can be utilized. If the test lead to the cell or battery is switched open and closed, the pointer will indicate the calibrating voltage value on the peak–to–peak scale. If the indication is incorrect, the AC calibration control (R26) is adjusted as required. As explained above, accidental overloads can change the values (or burn out) the AC multiplier resistors. In such a case, it is likely that the input blocking capacitor C1 will also have been damaged, and require replacement.

12–4 OSCILLOSCOPE TROUBLESHOOTING AND REPAIR

A simple oscilloscope has the typical appearance depicted in **Fig. 12–7.** Four general sections are employed, as shown in **Fig. 12–8.** If there is a

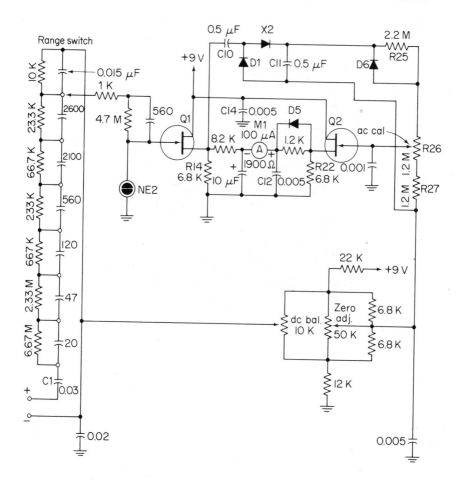

Figure 12–6 Configuration for the AC voltage–measuring function of a TVOM.

spot or trace displayed on the screen, both the low–voltage and the high–voltage power supplies are operating. In case the screen is dark, a technician proceeds to measure the power–supply voltages. Note that the high–voltage power supply in an oscilloscope can be very dangerous, because comparatively large values of filter capacitance are utilized. Therefore, an experienced technician always discharges the high–voltage filter capacitors before touching any part of the circuitry.

If the screen is dark, and the power–supply output voltages are normal, it is possible that the cathode–ray tube is defective. A substitution test is usually made. To anticipate subsequent discussion, a normal CRT will occasionally appear to be defective because of centering control–circuit defects that deflect the beam off–screen. On the other hand,

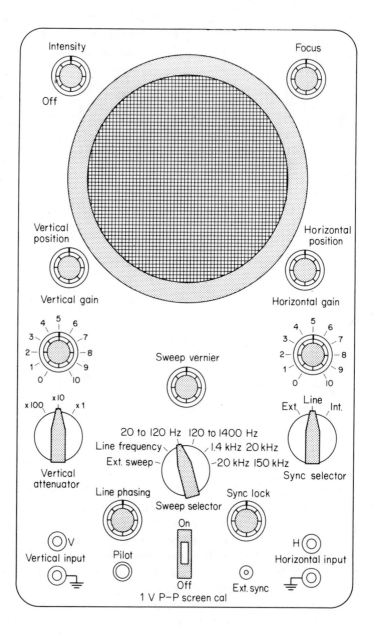

Figure 12–7 Appearance of a basic oscilloscope.

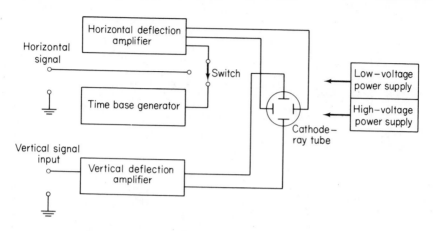

Figure 12-8 Block diagram of the basic oscilloscope.

if a spot or line is displayed on the screen, considerable preliminary information can usually be obtained by observing the action of the operating controls. As an illustration, if a horizontal line is displayed on the screen, and no vertical deflection is obtained when a finger is touched to the vertical-input terminal, there is trouble in the vertical-amplifier section. If a vertical line is displayed on the screen when a finger is touched to the vertical-input terminal, and no horizontal deflection is obtained by adjustment of the horizontal operating controls, there is trouble in either the horizontal-amplifier section or in the time-base section. To distinguish between these two possibilities, the scope may be set for external-sweep operation, and a finger then touched to the horizontal-input terminal. If a horizontal line then appears on the screen, the trouble will be found in the time-base section. Otherwise, the trouble will be found in the horizontal-amplifier section.

A vertical-amplifier configuration for a moderately elaborate oscilloscope is depicted in Fig. 12-9. DC coupling is employed, and the amplifier has full response down to zero frequency. Its upper frequency limit is approximately 5 MHz. High input impedance is provided by a field-effect transistor. This FET is protected against overload damage by diodes D423 and D424. The input section is arranged as a phase inverter, so that single-ended input provides push-pull output to the vertical-deflection plates of the CRT. A step attenuator is utilized for coarse gain control, and fine gain control is obtained by variation of the common-emitter load for Q464 and Q474. Control of vertical centering is provided by variation of the relative emitter bias voltages of Q464 and Q474.

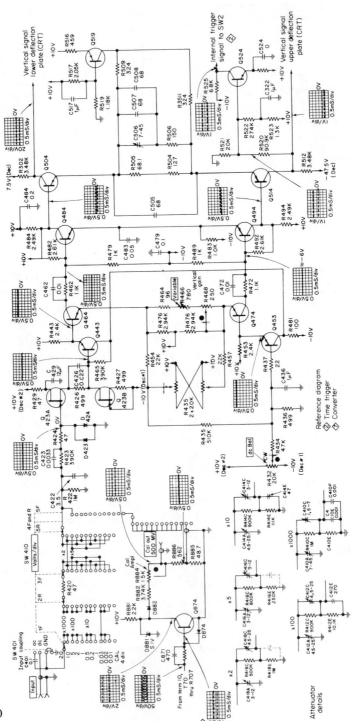

Figure 12–9 Vertical–amplifier configuration for a moderately elaborate oscilloscope (*Courtesy of Tektronix*).

If there is no output from the vertical amplifier, or if output appears at only one of the vertical–deflection plates, another scope should be employed to signal–trace the vertical amplifier. If the vertical amplifier is driven by an audio oscillator, for example, the sine–wave signal can be traced from the input terminal to the point at which the signal stops in the vertical amplifier. Then, DC–voltage measurements are usually most useful to close in on the defective component. Statistically, faulty capacitors are the most common culprits. As an illustration, if C422 becomes "shorted," the vertical amplifier will be "dead." Or if C436 becomes leaky, the beam may be thrown so far off–center that the centering control cannot bring the beam on–screen. Transistors occasionally become defective and require replacement. As explained previously, DC–voltage measurements, supplemented by low–power resistance measurements, are generally useful to evaluate the condition of a transistor.

Capacitor defects sometimes distort the frequency response of a vertical amplifier without producing other operating symptoms. For example, if C423 (Fig. 12–9) becomes "open," the high–frequency response of the amplifier will be poor. On the other hand, if C507 "shorts," the high–frequency response becomes excessive. Again, if C410F "opens up," square–wave reproduction will become distorted on the X1000 step of the vertical attenuator. To localize the cause of poor frequency response, the oscilloscope may be energized by a video–frequency signal generator, and the response of the vertical amplifier can be signal–traced with another oscilloscope. Note that the signal–tracing scope must have a frequency response that is equal to or greater than that of the scope under test.

Troubleshooting horizontal amplifiers entails the same general considerations as explained above. Some oscilloscopes have identical vertical and horizontal amplifiers. Other scopes provide limited–capability horizontal amplifiers and specialized vertical amplifiers. Lab–type scopes usually employ plug–in vertical and horizontal amplifiers, so that a wide range of test facilities can be provided economically. Most high–performance scopes have triggered time bases with calibrated sweeps. Note in Fig. 12–9 that the trigger signal is taken from the emitter of Q524. In turn, this sample of the vertical–amplifier signal is fed to the time–base trigger circuit depicted in Fig. 12–10. This circuit processes the incoming signal so that an output sync pulse is generated at a desired point along signal waveform.

As in the case of amplifier troubleshooting, localization of a faulty circuit branch in the time–base trigger section (Fig. 12–10) is accomplished to good advantage by using another scope as a signal tracer. Note that peak–to–peak voltage values should be measured, even though

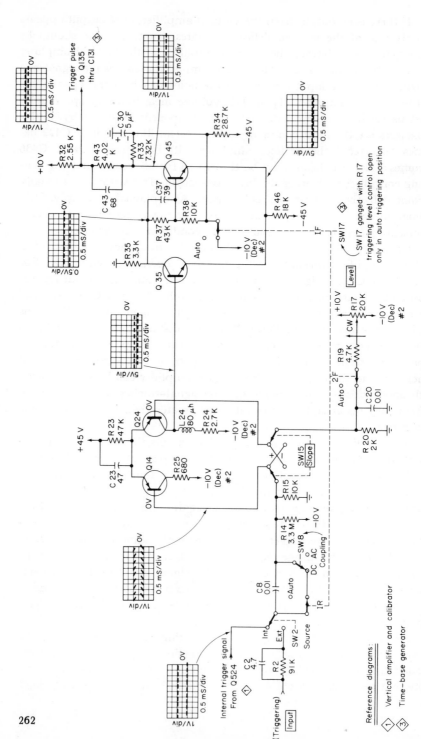

Figure 12–10 A time–base trigger circuit (*Courtesy of Tektronix*).

the signal waveform appears normal at a test point. That is, some component defects (such as a deteriorating transistor) can attenuate the signal substantially, although its waveshape is not greatly changed. The power–supply voltage is normally regulated, and no ripple is normally observed. In the event of power–supply trouble, a time–base trigger circuit may become erratic or inoperative due to incorrect supply voltage and/or excessive ripple voltage.

From the time–base trigger circuit, a trigger pulse is normally fed to the time–base generator circuit, exemplified in Fig. 12–11. In most trouble situations, the time base is "dead," but occasionally the time base operates continuously and does not respond to an applied trigger pulse. When the time base is inoperative, it is generally advisable to make a systematic check of the capacitors. As an illustration, if C131 is "open," the trigger pulse from Q45 cannot activate Q135. Note that in this situation, the time base will free-run in the Automatic mode at a rate of 50 Hz. Or, if C170 is "shorted," the time base becomes completely inoperative. In case there are no defective capacitors in the section, a leaky transistor or a diode with poor front-to-back ratio is likely to be causing the trouble symptom. Although resistors can become defective, this is a much less probable source of malfunction. In this regard, variable resistors, such as the stability control, may eventually become worn and erratic.

12–5 AM AND FM SIGNAL GENERATOR TROUBLESHOOTING

Various types of AM signal generators are used in electronics troubleshooting procedures. AM radio servicing requires a frequency coverage from approximately 100 kHz to 30 MHz. FM radio servicing requires a frequency coverage from 10 MHz to 108 MHz. Television servicing requires a frequency coverage from 4 MHz to 216 MHz; shops that service UHF tuners require generators with coverage up to 890 MHz. Fig. 12–12 shows a circuit diagram for a general–purpose wide–range oscillator with frequency coverage from 5 Hz to 1.2 MHz. Thus, this type of instrument serves both as an RF signal generator and an audio oscillator. It is essentially a stable variable–frequency oscillator followed by a step attenuator with a range of 80 dB in 10–dB steps.

Any signal generator should be checked at periodic intervals for accuracy of frequency calibration. That is, if a component such as a fixed capacitor, diode, or transistor starts to deteriorate, the first symptom that occurs is often a shift in frequency calibration. A secondary

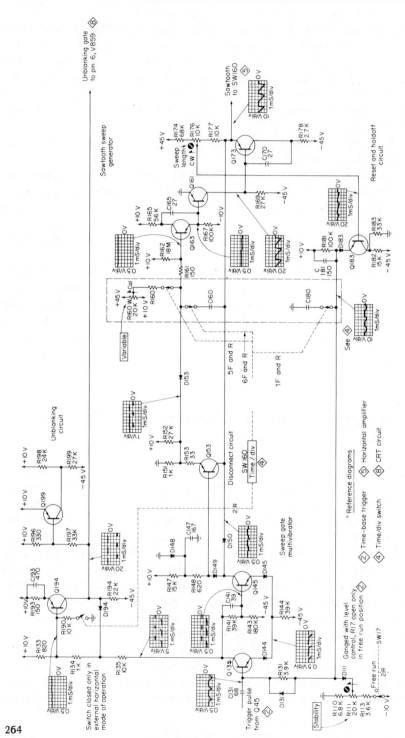

Figure 12–11 Configuration of a triggered time base (*Courtesy of Tektronix*).

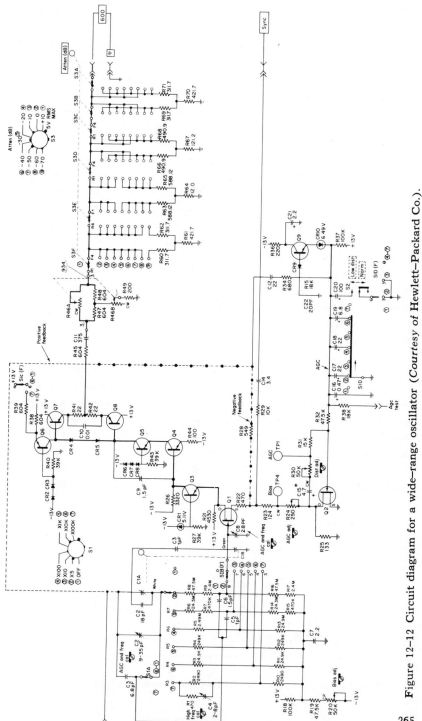

Figure 12–12 Circuit diagram for a wide-range oscillator (*Courtesy of Hewlett-Packard Co.*).

frequency standard, such as a quartz crystal oscillator, is commonly employed for spot–frequency checking. The outputs from the quartz crystal oscillator and the signal generator may be paralleled and fed into a signal tracer, for example. The generator tuning dial is then adjusted for zero-beat indication from the signal tracer, and the dial scale normally reads the same as the rated frequency of the quartz–crystal oscillator. Any error in dial calibration is corrected by adjustment of the frequency–calibration control(s) in the signal generator.

In case a substantial calibration error is found, it is good practice to check the oscillator circuitry in the generator. If a component or device has started to deteriorate, instrument operation is likely to become unreliable. Note that a calibration error may or may not be accompanied by a change in the rated output voltage level. As an illustration, if power–supply trouble occurs and the supply voltage is subnormal, both the calibration and the output level of the generator will be affected On the other hand, if a resistor in the oscillator circuit drifts in value, only the calibration accuracy will be affected, and the output level will remain unchanged. An off–value resistor or a defective switch contact in the attenuator section will cause an incorrect output level, but the frequency calibration will be unaffected.

An AM signal generator may employ an external modulator, or it may have built–in modulation facilities. In the example of Fig. 12–12, an external modulator is utilized. A typical modulator configuration is shown in Fig. 12–13. Amplitude modulation occurs in diode M, when a modulating voltage is applied. Note that a bias cell is provided to obtain optimum linearity of modulation. A small forward bias current flows through the diode and returns through the attenuator of the signal genator. Troubleshooting a modulator unit involves systematic component checks. For example, the modulating diode may develop a poor front–to–back ratio, or the blocking capacitor may become leaky or open.

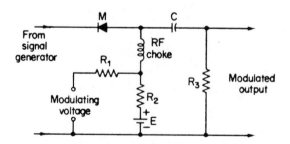

Figure 12–13 Example of an external modulator configuration.

FM signal generators and sweep–frequency generators are similar in many respects to AM signal generators. The chief difference is in the modulation mode. Another difference is found in the comparatively wide tolerance in calibration accuracy of an FM generator. The oscillator in an FM generator or sweep generator is usually frequency–modulated by means of a variable–permeability unit or a varactor diode. Figure 12–14 shows the plan of a variable–permeability modulator. This type of unit is comparatively long lived; however, capacitors may eventually become defective, or the bias rectifier may develop a poor front–to–back ratio. The configuration of a typical varactor–diode modulator is shown in Fig. 12–15. Malfunctions are most likely to result from capacitor faults, although active devices and varactor diodes eventually will also deteriorate. If an active device short–circuits, it is possible that one or more associated resistors will be damaged or destroyed.

12–6 TROUBLESHOOTING SQUARE–WAVE, PULSE, AND COLOR–BAR GENERATORS

Square–wave and pulse generators are basically multivibrator oscillators. A symmetrical multivibrator generates a square–wave output, whereas an asymmetrical multivibrator generates a pulse output. A color–bar generator contains several pulse generators and one or more square–wave generators which are utilized to form sync pulses, bar signals, and

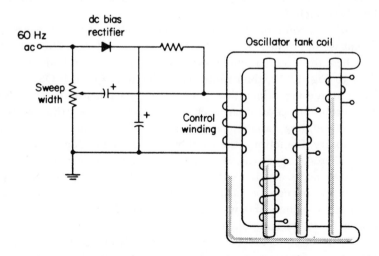

Figure 12–14 Plan of a variable–permeability (saturable reactor) type of FM modulator.

267

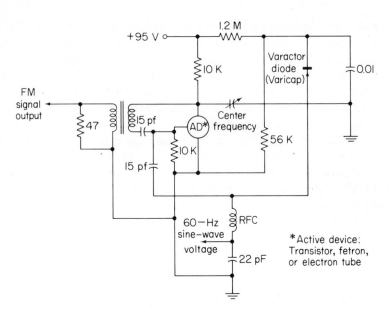

Figure 12–15 Example of a varactor–diode FM modulating arrangement.

crosshatch and dot patterns. Figure 12–16 exemplifies the configuration of a color–bar generator of the keyed–rainbow type. Three of the pulse generators employ blocking–oscillator circuits, and the others are of the RC design. Two quartz–crystal oscillators and an LC RF sine–wave oscillator are included.

One of the common trouble symptoms in this type of generating equipment is "scrambled" bar, crosshatch, or dot patterns. This difficulty occurs when the counter chain skips sync and triggers falsely. With reference to Fig. 12–16, the counter chain includes Q3, Q4, Q5, Q6, and Q7. Most triggering difficulties are caused by capacitor defects. Electrolytic decoupling capacitors, such as C1 and C2, are prime suspects. Leakage in coupling capacitors, such as C3, C4, C5, and C6, is also a good possibility. Sometimes the trouble area can be deduced from analysis of the distorted patterns. However, a systematic check of the capacitors becomes necessary in many situations. An oscilloscope and low-capacitance probe can be utilized to check the waveforms at various sections and branches, and waveform analysis will often assist in pinpointing a defective component.

Faulty transistors are localized to best advantage by DC voltage measurements. However, normal DC–voltage values may not be specified in the generator service manual. In such a case, it is often prac-

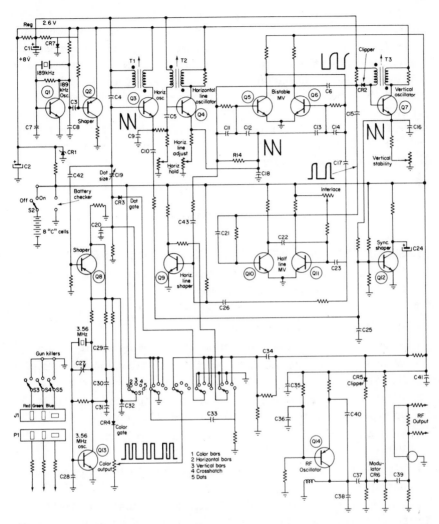

Figure 12–16 Configuration of a keyed–rainbow color–bar, dot, and crosshatch generator.

tical to make turn–off tests of suspected transistors. When analyses and tests are inconclusive, a substitution test is required. Semiconductor diodes occasionally become "open," "shorted," or develop a poor front–to–back ratio. A suspected diode should be disconnected at one end and checked with an ohmmeter. However, a zener diode is usually checked by substitution—if the diode measures an incorrect voltage drop in–circuit, it is most likely defective.

Blocking–oscillator transformers seldom become defective, and a faulty transformer is usually pinpointed by elimination of all other suspects. A substitution test is then made. Dirty or defective switch contacts can cause malfunctioning, particularly in generators that have been in long service. Resistors seldom cause trouble, and variable resistors are more likely to become defective than fixed resistors. Sometimes a transistor will become damaged or destroyed as a result of connecting a replacement battery in wrong polarity. If this accident occurs, the transistor(s) most likely to be damaged can be determined from an analysis of the generator circuit diagram.

QUESTIONS

1. Should VOMs and TVOMs be checked for calibration?
2. What would cause a VOM to change the zero setting when it is moved from one position to another?
3. How can the ohmmeter calibration be tested?
4. What is the accuracy of a typical ohmmeter?
5. How can the DC voltage indication of a VOM be accurately tested?
6. What is an accurate method of testing the AC voltage ranges of a VOM?
7. What is the factor that can be used to multiply the value of a DC voltage to obtain the correct value of the AC reading?
8. What are some of the problems that can impair the accuracy of a VOM?
9. What is the normal procedure when a VOM is found to be out of tolerance?
10. Why is the sensitivity of a TVOM greater than that of a VOM?
11. What is the first thing that you should check when the accuracy of the ohmmeter section is out of tolerance?
12. What clues does the presence of a spot on the screen give when troubleshooting an oscilloscope?
13. What safety precautions should a technician follow when testing a high voltage power supply?
14. What symptoms would you expect to observe on the screen of an oscilloscope in which the cathode–ray tube was defective?
15. Where is the trouble located if there is no time base in the internal-sweep position, but there is a horizontal line when the scope is placed

in the external–sweep position and the horizontal input touched with the fingertip?

16. How are the internal circuits of an oscilloscope signal traced?

17. What are some of the measurements that are used to evaluate the condition of a transistor?

18. What are the frequency requirements of an oscilloscope that is used to signal–trace a defective oscilloscope?

19. Why is another oscilloscope especially useful in signal–tracing the time–base trigger section?

20. What are the possible causes of an erratic or inoperative time–base generator?

21. Why should a signal generator be tested periodically?

22. What is the first symptom of possible failure of components such as capacitors, diodes or transistors?

23. How is error in dial calibration corrected?

24. How are AM signal generators generally modulated?

25. What is the chief difference between a sweep–frequency generator and an FM generator?

26. What is the basic circuit in a pulse or square–wave generator?

27. How should a possible defective diode be tested with an ohmmeter?

28. What are the most common problems in oscillators and generators?

13

MATV AND CATV
TROUBLESHOOTING

13-1 GENERAL CONSIDERATIONS

Master–antenna television (MATV) and community–antenna television
(CATV) systems are used in fringe areas or strong–interference areas in
order to provide good TV reception to all receivers connected into the
system. Although both systems have basic similarities, a CATV system
is comparatively complex and serves many more viewers. Figure 13–1
depicts a simple MATV installation serving three TV receivers. To obtain ample signal strength at each receiver, and to provide isolation
between receivers, a broadband VHF amplifier is utilized. Figure 13–2
depicts a more complex MATV installation with four single–channel
antennas and a distribution system to serve up to 100 TV receivers.

A balun is an impedance–transforming device; it usually matches
300–ohm and 75–ohm impedances. A hybrid is basically a signal–com-
bining device, or linear mixer. Distribution amplifiers supply each receiver with at least 1000 microvolts of signal. A splitter provides isolation between receivers so that the local oscillator of one receiver does not
produce interference with another receiver. In addition, a splitter may
provide amplification. The comparative complexity of a CATV system

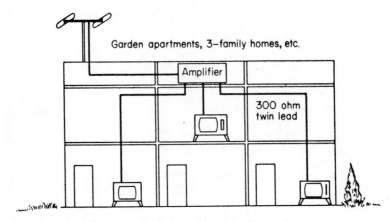

Figure 13-1 An MATV installation for a multiple dwelling.

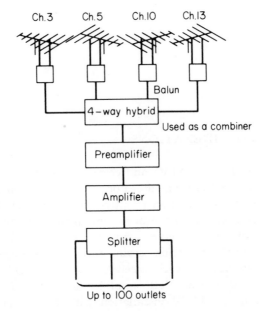

Figure 13-2 An apartment–building or motel MATV installation.

is seen in the block diagram of **Fig. 13-3.** An entire town or city may be served with **VHF, UHF, FM,** and locally produced programs.

A head end in a **CATV** system includes a **VHF** amplifier for each antenna, and a signal combiner. A translator heterodynes a **UHF** signal frequency into a **VHF** signal frequency. This is done to minimize the

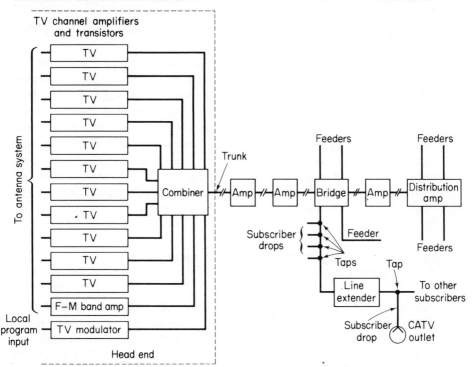

Figure 13–3 A simplified block diagram of a CATV system.

signal loss on long cable runs. A trunk cable conducts the signal energy from its originating site, such as on top of a mountain, to the utilization site, such as a town. Coaxial cable imposes a loss of about 1 dB per 100 feet. Thus, an amplification of approximately 50 dB per mile of cable is required. A bridger or bridging amplifier splits the signal into several portions for supplying feeder lines. Feeder amplifiers are also called line extenders. Both the main trunk amplifiers and the feeder amplifiers provide a gain up to 25 dB over a frequency range from 50 to 220 MHz. The system impedance is 75 ohms throughout.

Amplifiers and signal splitters are usually provided with a tilt control. This is a filter of the bandpass type with an insertion loss that is a function of frequency. The high end of the band is usually attenuated to some extent, because coaxial cable has a higher loss at higher frequencies. Strictly speaking, distribution cables are tapped off from feeder cables, but technicians often call both types distribution cables. Tap points are called subscriber taps. Note that a trunk line continues through a bridging amplifier. On the other hand, the last amplifier on a trunk cable is called a distribution amplifier. Feeder lines can be run up to 1000 feet from a bridger, before another amplifier (line extender) is required.

As seen in Fig. 13–3, a subscriber drop is a cable that connects to a tap point along a feeder line. An isolating resistor may be used between the subscriber drop and the feeder line. A small capacitor may be used instead, and is sometimes preferred because it introduces some tilt. Since distribution amplifiers must operate over a wide frequency range, they may be designed as separate high–band and low–band amplifiers, with a low–band tilt control and supply their outputs to a single line, as exemplified in Fig. 13–4. Class–A amplification is essential, because any amplitude nonlinearity will produce cross–modulation and mutual interference among the various signals.

13–2 TROUBLESHOOTING MATV AND CATV SYSTEMS

Customer relations are more of a problem in CATV servicing than in other types of electronic servicing. For example, if a subscriber reports trouble symptoms, the technician might need to check an amplifier or cables in a neighbor's backyard. Since the neighbor regards a technician as a nuisance in this situation, tact and diplomacy are prime requisites. It is essential for the technician to wear a uniform with his name in plain sight, and to carry an identification card. The ID card should be presented immediately when the doorbell is answered. After permission is obtained to enter the property, the technician must beware of dogs that might attack him. If possible, it is desirable to have the property owner standing by while the technician makes his tests and/or repairs. The following MATV and CATV troubleshooting procedures are typical.

Identification of Trouble. When a complaint of poor or no reception is called in, it is necessary to determine whether the trouble is being caused by a cable or a receiver defect. A quick check can be made by connecting the cable to another receiver that is in good operating condition. If the trouble is in the cable, the technician should first inspect the cable run from the receiver to the wall output. In many cases, dogs chew coaxial cables, causing short–circuits or open–circuits. Exposed cable runs in the backyard may also have been chewed by dogs. Underground cable runs may become defective due to leakage of water into the cable, or to mechanical damage. Common trouble symptoms caused by moisture are weak and snowy pictures, no high–band station reception, and weak or no color reception.

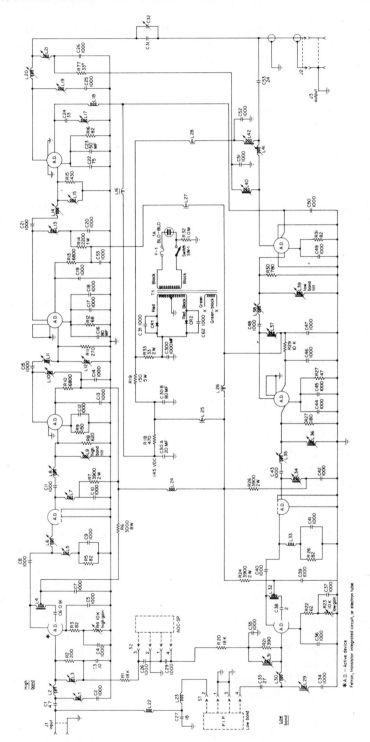

Figure 13–4 Configuration of a typical distribution amplifier.

Trouble Along the Right–of–Way. A common cause of outage in portions of a CATV system is due to car–pole accidents, particularly at night to trunk lines coming into sparsely populated towns. This type of damage is customarily repaired by the power company. Sometimes the power company may switch the circuits and erroneously leave the power supply for the cable system "dead." In these situations, it is difficult for the subscriber to understand why his television receiver does not work after the power line has been repaired. Again, this is an example of the requirement for good public relations.

Tampering by Subscribers. When a house is remodeled by a do–it–yourselfer, he may try to modify the cable installation. Sometimes the tyro will attempt to extend a cable run by means of ordinary twin lead, with the result that he obtains ignition noise and various forms of interference in the picture. If coaxial cable is utilized, a complaint of weak reception is made occasionally, due to the requirement for increased amplification. "Pirates," also called "bootleggers," sometimes attempt to evade subscriber fees by tapping into a cable run. In some designs, the cable also carries the supply voltage for the line extenders, and grounding of the cable by a "pirate" will blow the fuse and stop reception by all subscribers along one or more streets.

Sweep–Frequency Testing. Both troubleshooting and preventive maintenance are facilitated by sweep–frequency testing. A suitable generator for this type of test work is illustrated in Fig. 13–5. The generator signal is applied at the head end, and access points along the cable system are checked progressively with an oscilloscope. A scope man may have to work at amplifiers up in the air, with the aid of a "bucket" truck. If the frequency–response pattern shows excessive high–frequency or low–frequency attenuation, the equalizers or tilt controls may require realignment. When poor frequency response is being caused by reflections (standing waves), there is a cable fault that must be located. Most reflections are caused by water leakage into cables or boxes.

Amplifier Overload. An excessively high signal level is as undesirable as an excessively low level. If amplifier overload occurs, cross–modulation is the result. Typical overload symptoms are grain in the picture, herringbone interference, and the "windshield–wiper" effect (a weak interfering picture may appear with the desired picture, and the horizontal–blanking bar drifts back and forth across the image). Sometimes cross–modulation causes a weak interfering picture in which the vertical–blanking bar drifts up and down through the image. Therefore,

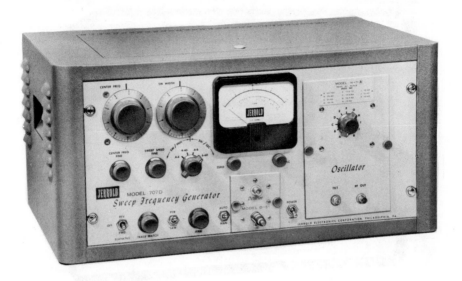

Figure 13–5 Wide–band sweep–frequency generator used to check CATV systems (*Courtesy of* Jerrold Electronics Corp.).

it is advisable for the technician to carry a field–strength meter along with a scope. (See Fig. 13–6.) If an abnormally high signal level is measured at an access point, the gain control of the associated amplifier should be adjusted accordingly.

Ghosts and Fuzzy Images. Ghosts and fuzzy images in receivers connected to a cable system are usually caused by a cable break or similar defect. However, some types of amplifier overload can simulate ghosts. Cable damage can be caused by construction crews who are unaware of buried cables, by vandalism, or by sabotage. Sometimes irresponsible boys with rifles shoot at CATV cables up in the air, and disgruntled subscribers have been known to chop cables with axes. Less serious image deterioration can result from "bootlegger" taps along a cable; these are sometimes difficult to find, unless a thorough and systematic checkout is made.

Hum in the Picture. In older types of MATV and CATV installations that employ tube–type amplifiers, 60–Hz hum bars occasionally appear in the picture. This symptom is usually caused by heater–

Figure 13–6 A field–strength meter used in MATV and CATV troubleshooting (*Courtesy of* Jerrold Electronics Corp.).

cathode leakage in an amplifier tube. Although hum bars are less commonly encountered when solid–state amplifiers are used, an open filter capacitor in the power supply will cause this trouble symptom.

Amplifier Troubleshooting. Troubleshooting amplifiers such as the one depicted in Fig. 13–4 involves the same basic principles that apply to RF and IF amplifiers in television receivers. However, more stringent specifications are placed on amplitude linearity in a CATV or MATV amplifier. Linearity can be accurately checked with a lab–type signal generator that has a calibrated output meter. As the signal level is varied through its normal range, the amplifier output should remain precisely proportional to the input level. Nonlinearity is generally caused by incorrect bias on a transistor, or by a defective transistor. Incorrect bias voltages are commonly caused by leaky capacitors.

QUESTIONS

1. What is a balun?
2. What is a splitter?

3. What units are usually included at the head end of a CATV system?
4. What is the purpose of a trunk cable?
5. What level of signal loss can be expected from a coaxial cable?
6. What is the system impedance of a CATV system?
7. What is a tilt control?
8. What is another name for tap points?
9. What is the maximum length of a feeder line before another amplifier must be employed?
10. What is the name given to the last amplifier on a trunk cable?
11. What is a subscriber drop?
12. What is the purpose of the small capacitor that is used between the feeder line and the subscriber drop?
13. What type of amplification is used in the CATV system?
14. What precaution should the technician take when repairing a CATV system in the yard of a property owner?
15. What is the first determination that a technician should make when there is a complaint of poor or no reception?
16. How can the technician quickly determine whether a problem is in the receiver or the cable?
17. What are some of the common causes of weak or no color reception?
18. What problems can the power company introduce into the system?
19. How are pirates discouraged in some CATV systems?
20. When is sweep–frequency testing used in a CATV system?
21. What causes most reflection (standing waves) in a CATV system?
22. What is the result if the signal is excessively high in amplitude?
23. What are typical symptoms of excessively high level of signal?
24. What causes ghosts or fuzzy images in the system?
25. What is the probable cause of 60–Hz hum in older systems?

14

MARINE ELECTRONIC TROUBLESHOOTING

14-1 GENERAL CONSIDERATIONS

Marine electronic equipment is extensive, ranging from communications receivers and transmitters through various specialized units such as depth sounders, fuel vapor detectors, foghorn hailers, fish spotters, tachometers, and so on. Troubleshooting marine radio receivers and transmitters involves the same basic procedures that have been explained for land–based equipment. Therefore, this chapter covers specialized marine electronic equipment. Note that test equipment used in marine electronic servicing should be portable and battery–powered. The same types of test equipment used in land–based radio and television troubleshooting are utilized in marine work. For example, a VOM or TVM is basic, and an AM/FM signal generator covering the communications bands is often required. An audio oscillator with output in the ultrasonic frequency range is useful, and a small general–purpose oscilloscope is very helpful. A kit of replacement transistors and electronic components will be needed, with typical items noted in the following pages.

14-2 DEPTH-SOUNDER TROUBLESHOOTING

In order to analyze trouble symptoms in a depth sounder, it is necessary to understand its functions. With reference to Fig. 14–1, we observe the time relationships of the ultrasonic sound waves in the water. A transducer is employed, to which electrical pulses are applied. Inside the transducer, a piezoelectric ceramic element oscillates at approximately 200 kHz, and vibrates in much the same way as the cone of a loudspeaker. This vibration produces ultrasonic waves directed toward the bottom of the water, in a comparatively narrow beam. Because these waves are not audible, they do not produce any sound although they travel at the rate of sound through the water at approximately 4800 feet per second, or 0.0002083 second per foot. As each pulse of ultrasonic waves leaves the transducer, it causes a neon lamp to flash at the zero depth mark on the scale depicted in **Fig. 14–2**.

The time required for the ultrasonic wave, or signal, to travel from the transducer to the bottom is equal to the product of water depth times the speed of sound in water. As an illustration, if the depth of water under the boat is 30 feet, then 30 $\times$ 0.0002083 or 0.00625 second of

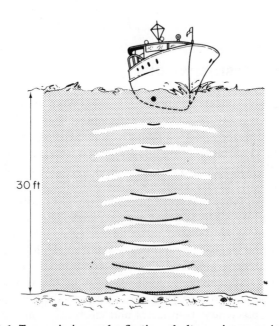

30 ft

Figure 14–1 Transmission and reflection of ultrasonic waves in water.

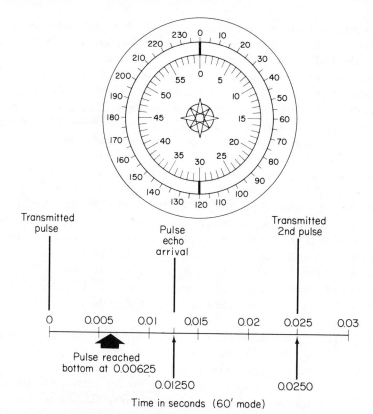

Figure 14–2 A depth indication of 30 feet on the inner scale.

sound–travel time is involved. Next, as the signal strikes the bottom, part of it is reflected back up to the transducer at the same speed. The return time for the reflected signal is another 0.00625 second, making the total travel time in this example 0.01250 second. Upon arriving at the transducer, the reflected signal energy causes the ceramic element to vibrate at the same frequency that was originally generated. This vibration causes the element to produce a pulsed 200–kHz signal that is then amplified and applied to the neon lamp, causing it to flash.

In Fig. 14–2, with the unit operating in the 60–foot mode, a nylon disc is spinning at the rate of 2400 rpm. This speed represents a time of 0.0250 second for each revolution. In turn, the nylon disc would have rotated a half revolution during the signal time of 0.01250 second. Thus, the neon lamp would flash at the bottom of the inner scale, at the 30–foot mark. Again, if operated in the 240–foot mode, the nylon disc spins at the rate of 600 rpm. This speed represents a time of 0.10000 second for each revolution. In turn, the nylon disc would have rotated only ⅛ of one revolution during the signal time of 0.01250 second. Thus, the neon lamp would flash at the 30–foot mark on the outer scale.

Since the transmit–receive cycle is repeated 40 times per second in the 60–foot mode, or 10 times per second in the 240–foot mode, the practical result is equivalent to a continuous sounding. Therefore even relatively small variations of the bottom contour can be observed while the boat is in motion. A block diagram of the exemplified depth sounder is shown in Fig. 14–3. For details of the various blocks, refer to the schematic diagram depicted in Fig. 14–4. The nylon disc is rotated at a controlled speed by the motor. For each revolution of the disc, the permanent magnet mounted on its face will pass over pickup coil L1 and induce a voltage pulse in the coil. This voltage pulse is applied through R7 to the base of Q4, thus causing Q4 to conduct and produce a negative pulse at its collector. The negative pulse is then differentiated by C4 and R11 (also by R12 in the 240 mode) and applied to Q5. This causes Q5 to turn off and produce a negative pulse at its collector. Note that Q5 and its circuitry provide a delay to match the zero flash on the dial with the zero markings.

The negative pulse from the collector of Q5 is differentiated by C5 and R15 (also by R16 in the 240 mode) and applied to Q6. This causes Q6 to turn off and produce a negative pulse at its collector. Q6 and its circuitry function much like Q5, but the Q6 circuit determines the time duration of the transmit pulse. The negative pulse taken from the collector of Q6 is then applied to the base of Q7, which causes Q7 to conduct and produce a positive pulse at its collector. This positive pulse is applied to the base of Q8 which, because of the tuned circuit comprised of coil L2, capacitor C10, and the transducer, generates an oscillatory signal of approximately 200 kHz during the time of the applied pulse. Capacitor C8 couples regenerative feedback to the base of Q8, and with C9 determines the necessary amount of regeneration to sustain oscillation.

In turn, the 200–kHz signal is increased in amplitude to approximately 200 volts peak-to-peak by L2, and is then applied to the transducer. Accordingly, the transducer converts the electrical signal to an ultrasonic signal which is coupled into the water. (See Fig. 14–5). Diodes D2 and D3 serve to suppress any strong ringing oscillations developed by L2. The foregoing circuit actions occur only at the time that the magnet induces voltage into the pickup coil. Consequently, the transducer can generate an ultrasonic signal only during this same time. In between transmit periods, the transducer waits to receive any signals reflected from the bottom.

The negative pulse produced at the collector of Q4, as described

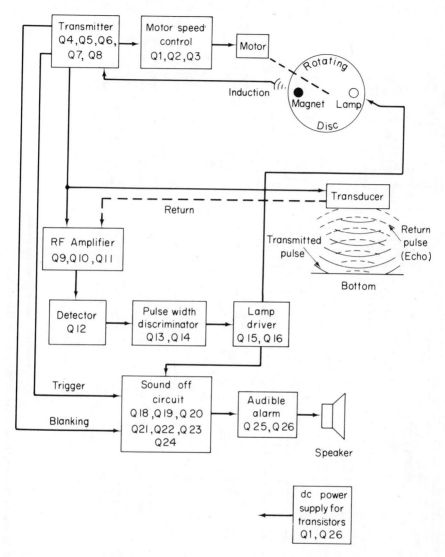

Figure 14–3 Block diagram of depth sounder (*Courtesy of* Heath Co.).

previously, is also applied to the motor–speed–control circuits. Here, the negative pulse is differentiated by C2, R1, and R2 (and also R3 and R4 in the 240 mode), and applied to the base of Q1. Thereby, Q1 is momentarily turned off to produce a negative pulse at its collector. The duration of this pulse, which is controlled by the differentiating com-

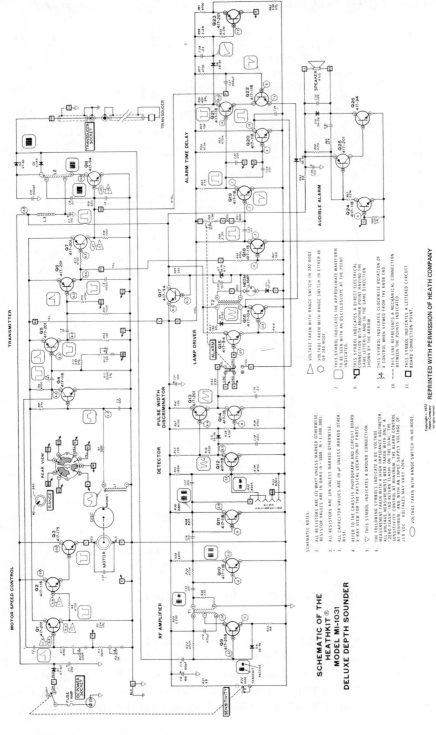

SCHEMATIC OF THE
HEATHKIT®
MODEL MI-1031
DELUXE DEPTH SOUNDER

Figure 14–4 Schematic diagram of the depth sounder (*Courtesy of Heath Co.*).

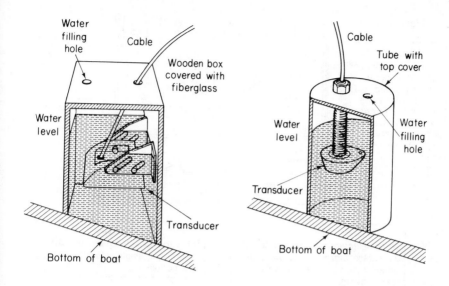

Figure 14–5 Typical transducer mounting arrangements (*Courtesy of* Heath Co.).

ponents, can be varied by means of trimmer resistor R2 in the 60 mode, or by trimmer resistors R2 and R4 in the 240 mode. The negative pulse taken from the collector of Q1 is then applied to the base of Q2, and turns off both Q2 and Q3 for a short time. When Q3 is turned off, no power is applied to the motor.

All of the circuit actions described above occur only at the time that the magnet induces voltage into the pickup coil. Therefore the motor is turned off for a period of time during each disc revolution in order to control the motor speed. Note that the receiver circuits operate continuously whenever the depth sounder is operating. The ultrasonic signal produced by the transducer, when reflected from the bottom or from a submerged object, returns to the transducer, where it is changed back into an electrical signal. This return signal is coupled via C13 to the base of Q9. Diode D4 limits the amplitude of the signal to prevent damage to Q9, while the sensitivity control R22 determines the gain of that stage. (See Fig. 14–6.) From the collector of Q9, the signal is applied to the primary winding of transformer T1. Note that T1 and C14 form a tuned tank circuit (bandpass filter) that responds only to those frequencies near 200 kHz. This optimizes the signal–to–noise ratio.

From the secondary of T1, the signal is coupled to Q10 and Q11 for further amplification and to the base of Q12. Note that Q12 forms a de-

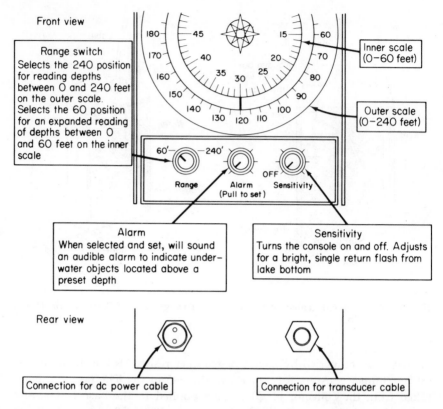

Front view

Range switch
Selects the 240 position
for reading depths
between 0 and 240 feet
on the outer scale.
Selects the 60 position
for an expanded reading
of depths between 0
and 60 feet on the inner
scale

Inner scale
(0–60 feet)

Outer scale
(0–240 feet)

Range   Alarm   Sensitivity
              (Pull to set)

Alarm
When selected and set, will sound
an audible alarm to indicate under-
water objects located above a
preset depth

Sensitivity
Turns the console on and off. Adjusts
for a bright, single return flash from
lake bottom

Rear view

Connection for dc power cable

Connection for transducer cable

Figure 14–6 Control functions of the exemplified depth sounder (*Courtesy of* Heath Co.).

tector circuit with R36 and C18, which changes the signal into a negative pulse. This pulse is coupled via D5 to turn off Q14, and at the same
time is coupled via R36 to turn on Q13. Thereby, C19 charges at a rate
determined by R41 until it reaches approximately 1.2 volts. This allows
Q15 and Q16, which are normally biased off, to conduct. The resulting
output signal from Q16 is inductively coupled through T2 where its amplitude is increased sufficiently to fire the neon lamp. Diode D6 prevents
the collapsing magnetic field of T2 from firing the neon lamp a second
time, which would cause two flashes for each pulse. Diode D11 prevents
the lamp from firing in a reverse direction.

At the end of a return signal pulse, Q12 returns to cutoff. In turn,
its collector voltage increases and Q14 goes into conduction. This provides a discharge path for C19. Received noise pulses, even though they
may have a high amplitude, are usually too narrow to allow C19 to

charge enough to turn on Q15 and Q16. False flashing of the neon lamp is thus avoided. The negative pulse at the collector of Q6, as explained previously, is also coupled through R46 to the base of Q18. This causes Q18 to momentarily turn off and produce a positive pulse at its collector. From the collector of Q18, the pulse is coupled through C23 to the base of Q19. At this point, the pulse width can be varied by the depth–alarm control, R49. From the collector of Q19, the pulse is coupled through C24 (also C25 in the 240 mode) to the base of Q20. When the alarm control is pulled to the out position, Q20 turns on Q15 and Q16 to allow the neon lamp to flash a display of the depth at which the alarm is set.

The pulse at the collector of Q19 is also coupled through R53 to the base of Q21 and causes Q21 to turn on. However, no signal will appear at the collector of Q21 unless Q22 is also turned on. Q22 has an RC network in its base circuit (R57 and C27). This network holds Q22 cut off for a short period of time after each transmit pulse. When Q21 and Q22 conduct at the same time, the signal at the collector of Q21 causes D9 to conduct, and C28 charges. The charge on C28 turns on Q23 and Q24. The resulting signal at the collector of Q24 causes the multivibrator circuit of Q25 and Q26 to actuate the speaker and produces an audible alarm.

Troubleshooting a depth sounder generally involves preliminary localization. If either the receiver or the transmitter section has a defect, the system fails. An oscilloscope is the most useful signal tracer, although an AC TVM will serve the purpose, if necessary. Normal operating waveforms are depicted in Fig. 14–4, for the equipment exemplified. If the system is completely inoperative, it is advisable to check the supply voltage. Then, if other possibilities must be checked, try a new fuse, test diode D1, and make certain that the on–off power switch is not defective. Note that if a new fuse blows immediately, there is a short–circuit that must be located and cleared. A short–circuited capacitor is the most likely suspect.

If signal–tracing tests localize the trouble in a particular stage, it is usually advisable to make DC–voltage measurements in the trouble area, and to compare the measured values with those specified on the schematic diagram. An incorrect DC voltage value at a transistor terminal generally indicates either a defective transistor or a leaky capacitor in the associated circuitry. It is sometimes possible to distinguish between these two possibilities by turning the equipment off and making circuit–resistance measurements with a hi–lo pwr ohmmeter. Off–value resistors can often be identified using the lo–pwr ohms function. It is sometimes helpful also to make turn–off and turn–on tests of suspected transistors.

If the signal channels operate normally, but the disc does not turn, transistors Q1, Q2, Q3, Q4, and Q17 should be checked. The same trouble symptom can be caused by a defective diode D7. In rare cases, the motor may be found defective. Next, if the disc turns, but the lamp will not flash, coil L1 may be misadjusted or defective. Another possibility is poor contact of a brush to a slip ring, or a magnet out of position. Transistors Q5 through Q17 should be checked, and, if they are normal, diodes D2 through D6 also fall under suspicion. Note that the neon lamp eventually requires replacement. A fault in L2, T1, or T2 can also cause this trouble symptom.

If the lamp flashes normally at zero, but there is little or no return flash, remember that the water depth might exceed 240 feet, or there might be a very soft bottom that absorbs an excessive amount of the signal. If the motor speed cannot be controlled, transistors Q1 through Q4 should be checked. This difficulty can also be caused by a defect in L1, or by an improperly positioned magnet. In case the audible alarm does not work, transistors Q18 through Q26 should be checked. Diodes D8, D9, and D10 should also be checked. Fixed capacitors are ready suspects in equipment that has been in service for an extended length of time. The speaker might become defective, although this is less likely.

If bright flashes occur over the entire range of the dial, the most likely cause is an excessively high setting of the sensitivity control. Otherwise it is advisable to check out transistors Q6 and Q13. Another possibility is an incompletely submerged transducer. Note also that ship–to–shore radiotelephone transmissions tend to radiate a large amount of RF energy. If these transmissions originate near the depth–sounder location, they may produce interference. Still another possibility is a failure in some portion of the ignition–noise–suppression system.

14–3 FUEL–VAPOR DETECTOR
TROUBLESHOOTING

A fuel–vapor detector indicates the presence of gasoline vapor in the bilge, which could cause an explosion. With reference to Fig. 14–7, the arrangement consists of a sensing element, control unit, and interconnecting cable. The element is formed from a small platinum–wire filament housed in a glass tube. This tube has openings at the top and bottom to let air pass through. It is placed in a stainless–steel mesh shield. Air passes through the shield and glass tube, past the filament.

In operation, the wire filament is heated by current flowing through it from the control unit. If an explosive vapor comes into contact with

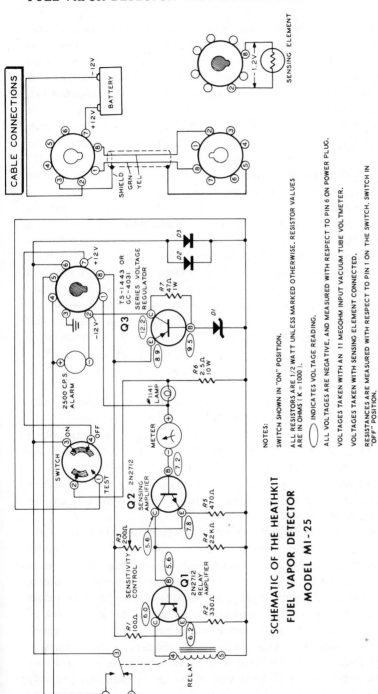

Figure 14–7 Circuit of fuel–vapor detector (*Courtesy of Heath Co.*).

the sensing element, a chemical reaction occurs which causes the temperature of the filament to increase. In turn, the filament resistance increases, and the voltage drop across the filament changes. This change is applied to the control unit through the interconnecting cable. The control unit is composed of a relay amplifier Q1, sensing amplifier Q2, and series voltage regulator Q3. When the switch is turned on, a positive voltage is applied from the battery through D2 and D3 to the voltage regulator circuit, consisting of Q3, R7, and D1. This circuit regulates the voltage to Q1 and Q2 and to the sensing element. Diodes D2 and D3 prevent the circuit components from being damaged in case the battery voltage might accidentally be reversed.

Transistors Q1 and Q2 operate as a two–stage trigger circuit. In normal operation (when an explosive vapor is not present at the sensing element), Q1 conducts more than Q2. When Q1 conducts, current flows through the relay winding and opens the relay contacts, disconnecting the battery voltage to the alarm and any external warning device. On the other hand, when an explosive vapor is present at the sensing element, the voltage drop across the element increases. The base of Q2 becomes less negative than the emitter. This causes Q2 to conduct, and the meter deflects. At the same time, the emitter of Q1 becomes more positive than the base. In turn, no current flows through the emitter-collector circuit and the relay contacts close, connecting the alarm circuit to the battery.

When the switch is turned to the Test position, R6 is shorted out, and Q2 then conducts more than Q1. This simulates an explosive condition, in order to check the control–unit circuits and to clean the oxide from the filament of the sensing element. If the result of the test is unsatisfactory, the battery supply voltage should be checked first. In case the supply voltage is normal, the circuit voltage and resistance values should be measured. Figure 14–8 shows voltage and resistance charts for the configuration of Fig. 14–7.

If the pilot lamp fails to light, it may be burned out. Other possibilities are a defective function switch, or R6 open–circuited. If the meter goes full–scale but the pilot lamp fails to light, the sensing element may be open, or there may be an open–circuit in the cable between the control head and the sensing element. If the pilot lamp lights but the meter does not deflect and the sensitivity control does not work, the meter may be burned out, transistor Q2 may be defective, or control R3 may be open.

If the meter indication changes when the battery voltage changes, transistor Q3 might be open–circuited, R7 could be open, or the Zener diode may be short–circuited. Note that if the meter indicates properly but the external alarm does not function, Q1 is possibly defective, or the

VOLTAGE CHART

	B	E	C
Q1	5.6	6.2	6.0
Q2	7.2	7.8	5.6
Q3	9.5	8.9	12.2

RESISTANCE CHART

	B	E	C
Q1	800 Ω	80 Ω	620 Ω
Q2	530 Ω	62 Ω	800 Ω
Q3	5 Ω	660 Ω	620 Ω

All voltages are negative, and measured with respect to pin 6 on the power plug.

Voltages taken with an 11 megohm input vacuum tube voltmeter.

Voltages taken with sensing element connected.

Resistances are measured with respect to pin 1 on the switch. Switch in 'off' position.

Figure 14—8 Voltage and resistance charts.

relay contacts might be bent, dirty, or burnt. If the meter starts to indicate toward "dangerous" after the unit has been turned on for some time, Q3 may be defective, or the Zener diode may have become unstable.

14—4 FOGHORN HAILER TROUBLESHOOTING

Figure 14—9 shows the schematic diagram for a foghorn hailer that operates as a combination foghorn, hailer, and listener. When used with accessory speakers, it also serves as an intercom. It provides a foghorn signal that can be heard a mile away. Duration and rate of the signal are adjustable. Between foghorn blasts, the circuits automatically return to the listen mode. The boat horn mode produces a higher pitched signal which is used to warn other craft or to signal bridges. In the hail mode, a switch is pressed which changes from the listen circuit to the call circuit. The operator's voice is projected for several hundred yards, for use when coming in to dock or calling other craft.

The circuit can be subdivided into a tone—generator section, multivibrator, relay, and amplifier section. (See Fig. 14—10.) The tone generator produces the low and high pitched signals for the foghorn and boat horn sounds. This generator is a phase—shift oscillator consisting of Q1 and its associated resistors and capacitors. Three RC combinations are used for phase shifters in the feedback loop. Each combination contributes approximately 60 degrees in phase, or a total of 180 degrees to

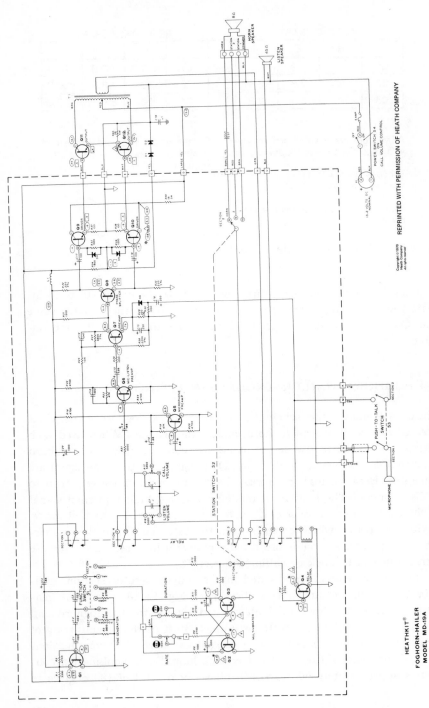

Figure 14-9 Foghorn hailer schematic diagram (*Courtesy of Heath Co.*).

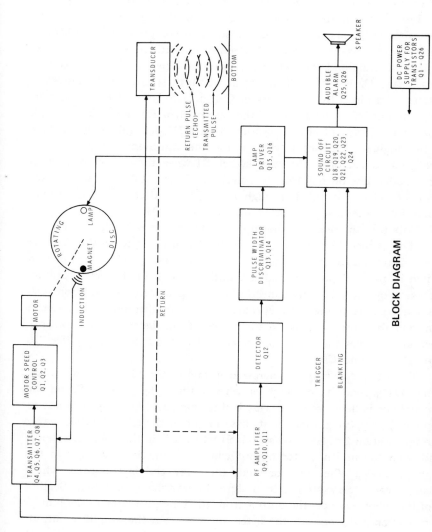

BLOCK DIAGRAM

Figure 14–10 Block diagrams for the foghorn hailer (*Courtesy of Heathkit*).

sustain oscillation. These combinations are C and the input impedance of Q1; C2 and R3; and C3 with either R4 or R5. The output frequency of the oscillator depends on the values of the resistors and capacitors in the circuit. The selection of R4 or R5 by S1 determines the different pitches. Note that the output from the oscillator is coupled through C4, section 1 of the relay, and R25, to the base of amplifier transistor Q7.

The multivibrator controls the on and off durations of the foghorn. It consists of Q2 and Q3 and their associated components in an astable multivibrator circuit. This has two stable conditions. In one condition Q2 is conducting and Q3 is cut off, and in the other condition Q3 is conducting and Q2 is cut off. The rate at which these two conditions alternate is determined by the values of C5 and C6, R9 and R10, and controls R6 and R7. Adjustment of the rate control R6 determines the off time, and adjustment of the duration control R7 determines the on time of the foghorn. Operating voltage for the multivibrator is supplied through section 2 of the function switch S1 when the unit is in the foghorn mode. The output from the multivibrator is taken from the collector of Q3 and coupled through R12, R14, and section 1 of station switch S2 to the base of relay control transistor Q4.

Note that the relay, as shown in the schematic, is in its resting position. This places the foghorn hailer in the listen mode when the function switch is turned to its hail position. (See Fig. 14–11.) With the function switch in the hail position and the station switch in the horn position, section 1 of the relay disconnects the tone generator output from Q7. Section 2 connects the listen speaker to the amplifier output. Section 3 connects the horn speaker to the wiper of the listen volume control. Section 4 couples the listen volume control to the listen preamplifier Q6 through R21 and C9.

The relay coil may be energized in one of the following two ways. When Q4 is made to conduct, or when push–to–talk switch S3 on the microphone is depressed, the relay coil is energized. Q4 may be made to conduct by applying a bias to its base in one of the following two ways: through section 2 of function switch S1 when it is in the boat horn mode, and from the multivibrator output when the function switch is in the foghorn mode. When Q4 conducts, the relay coil is energized, which changes each relay section to a different switching condition. Section 1 of the relay connects the output from the tone generator to the amplifier. Section 2 disconnects the listen speaker from the amplifier output. Section 3 connects the horn speaker to the amplifier output. Section 4 can be disregarded when the function switch is in the horn or foghorn mode. The signal from the tone generator is connected to the horn speaker through the amplifier when the function switch is depressed.

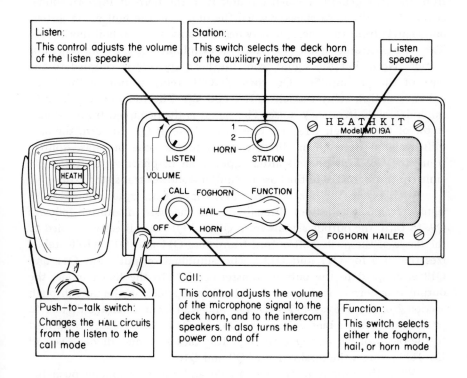

Listen:
This control adjusts the volume of the listen speaker

Station:
This switch selects the deck horn or the auxiliary intercom speakers

Listen speaker

LISTEN   STATION

VOLUME

CALL FOGHORN   FUNCTION

HAIL

OFF   HORN

HEATHKIT
Model MD 19A

FOGHORN HAILER

Push–to–talk switch:
Changes the HAIL circuits from the listen to the call mode

Call:
This control adjusts the volume of the microphone signal to the deck horn, and to the intercom speakers. It also turns the power on and off

Function:
This switch selects either the foghorn, hail, or horn mode

Figure 14–11 Controls and functions (*Courtesy of* Heathkit).

When the function switch S1 is turned to the foghorn mode, section 1 selects R5 for the foghorn tone, and section 2 connects the supply voltage to the multivibrator.

When Q3 of the multivibrator conducts, its collector voltage is so low that Q4 will not conduct. The relay will then remain in its resting position, which causes the foghorn hailer to operate in its listen mode. When the multivibrator changes its state, Q3 cuts off and its collector voltage increases, causing Q4 to conduct, which energizes the relay coil. This causes the relay switch sections to be in the same condition that they are in for the boat horn mode. The foghorn signal is then coupled through the amplifier to the horn speaker. When the multivibrator again changes its state and Q3 conducts, the foghorn hailer again returns to the listen mode. These alternating periods of sound and silence continue as long as the function switch is left in its foghorn position.

Note that the amplifier will process three different signals: the signal from the tone generator when the unit is in the horn or foghorn mode; signals from the horn speaker when the unit is in the hail mode, listen; and signals from the microphone when the unit is in the hail mode, call. The tone generator signal to Q7 has been explained in the tone generator section. Note that the signal from the horn speaker is coupled to the base of the preamplifier, Q6, through Q21, relay section 3, the listen volume control, relay section 4, R4, and C9. The function of C7 is to minimize high–frequency noise. The signal is then coupled from the collector of Q6 through C13 and R26 to the base of Q7. In turn, the amplified signal from Q7 is direct–coupled to the base of phase–splitter transistor Q8. This phase splitter supplies an in–phase signal (from its emitter) through C18 to the base of emitter–follower transistor Q10.

At this time, the phase splitter also supplies an equal out–of–phase signal (from its collector) through C17 to the base of emitter–follower Q9. In turn, the output signals from Q9 and Q10 are directly coupled to the bases of the push–pull output transistors Q11 and Q12. (R40 stabilizes the bias for Q11 and Q12.) The amplified output from Q11 and Q12 is coupled to the output transformer T1. In the listen mode, the output transformer is connected to the listen speaker in the unit through section 2 of the relay. Diodes D1, D2, and R34 form a voltage divider to provide the proper bias voltage for Q9 and Q10. This voltage is applied to the bases of Q9 and Q10 through R35 and R36 and diodes D4 and D5. R39 reduces the dissipation of Q9 and Q10.

When the unit is in the hail mode and the microphone push–to–talk switch S3 is depressed, the relay coil is energized through section 2 of S3. At the same time the microphone is connected into the system through section 1 of switch S3. The microphone signal is coupled to the microphone preamplifier Q5 through C11. This signal is then coupled through the call volume control, section 4 of the relay, R21 and C9 to the base of Q6. With the relay energized, the listen speaker is disconnected from the output transformer, section 2 of the relay, and the horn speaker is connected to the output transformer through section 3 of the relay. The horn speaker serves as both loudspeaker and microphone.

If auxiliary speakers are used for an intercom system, the station switch S2 is used to select the horn speaker or an auxiliary speaker. Section 2 selects the speaker, and section 1 disconnects the input voltage from relay control transistor Q4 to prevent the horn or foghorn from accidentally sounding through an auxiliary speaker. Whenever the push–to–talk switch on the microphone is released, a high voltage is generated by the collapsing magnetic field in the relay coil. This momentary high voltage could cause a popping sound in the listen speaker. However, this voltage is coupled to Q7 via C16 and D3 as a degenerative feedback voltage which momentarily biases off Q7 and eliminates any popping sound.

C21 is a high–pass filter which eliminates the possibility of applying full power to the horn at frequencies lower than the speaker is designed to handle. C12 and R23 in the Q6 circuit and C14 and R27 in the Q27 circuit form high–frequency degenerative feedback loops which attenuate the high frequencies. This attenuation of the high frequencies provides a powerful penetrating call signal without annoying harshness.

If the unit is completely inoperative, fuse F1 should be checked first, and the supply voltage verified. If no click is heard when the microphone button is depressed, the supply voltage may be low, transistor Q5 may be defective, or C9 might be faulty. If the boat horn does not sound and the relay does not click, the function switch may not have been set to the horn position. Otherwise, transistors Q2, Q3, and Q4 should be checked. If the relay clicks but the boat horn does not sound, the power source may be weak, Q1 or C4 might be defective, or the function switch may have become defective. If the foghorn sounds but will not cycle automatically, or if the rate and duration adjustments are inoperative, transistors Q2 and Q3 should be checked.

In some cases the relay clicks when power is applied, but will not drop out, and the unit is locked in the hail position. The most likely suspects in this situation are C16 and Q4. If one horn (boat horn or foghorn) works, but the other does not, it is likely that R4 or R5 is burned out. Note that if the fuse blows when the unit is turned on, C20, Q11, Q12, T1, and the insulators at Q11 and Q12 should be checked. When the foghorn duration or rate is too long or too short, the supply voltage may be abnormal or subnormal, or C5 or C6 may be defective. If a pop is heard in the unit speaker when the function switch is returned from the hail position, the most likely suspects are C14, C15, and D3.

If the sound does not stop normally at the end of the foghorn cycle, D3, C15, and C16 should be checked. Ignition noise in the speaker points to a defect in C20 or L1, or to the possibility that the cabinet may be poorly grounded. Weak horn and foghorn outputs are generally caused by a subnormal supply voltage. If the hail or listen signal is distorted, D1 and D2 should be checked. When analysis of circuit action is required to close in on a defective component, DC–voltage measurements are generally most informative. The measured values should be compared with the specified values on the schematic diagram. Resistance measurements with a hi–lo pwr ohmmeter will often provide additional useful data.

QUESTIONS

1. What are two basic requirements of test equipment used in marine electronics?
2. What are some of the basic test instruments necessary for the maintenance of marine electronics equipment?
3. What is the basic requirement for repairing any electronics equipment?
4. How does the ceramic element in the depth sounder produce signals?
5. What type of signals does a depth sounder transmit?
6. What determines the repetition of the transmitted signals from the transducer of the depth sounder?
7. When does the receiver circuit operate in a depth sounder?
8. In Fig. 14–6, what is the purpose of the tuned tank circuit made up of T1 and C14?
9. In Fig. 14–6, what is the circuit action when the alarm control is pulled to the out position?
10. What type of alarm does the unit produce?
11. What is one of the problems in troubleshooting a depth sounder?
12. How may suspected transistors be tested?
13. What is one of the common causes of bright noises occurring over the entire range of the depth sounder?
14. What is the function of a vapor detector?
15. How does the vapor detector sense the presence of an explosive vapor?
16. How is the vapor detector tested for operation?
17. What are the symptoms of an inoperable sensitivity control?
18. How is the foghorn used as an intercom?
19. What are the base circuits of the foghorn hailer in Fig 14–10?
20. In Fig. 14–11, how is the relay coil energized?

INDEX